T0234510

The
Chemistry
Companion

The
Chemistry
Companion

A C FISCHER-CRIPPS

CRC Press
Taylor & Francis Group
Boca Raton London New York

CRC Press is an imprint of the
Taylor & Francis Group, an **informa** business

CRC Press
Taylor & Francis Group
6000 Broken Sound Parkway NW, Suite 300
Boca Raton, FL 33487-2742

© 2012 by Taylor & Francis Group, LLC
CRC Press is an imprint of Taylor & Francis Group, an Informa business

No claim to original U.S. Government works

Printed in the United States of America on acid-free paper
Version Date: 20110517

International Standard Book Number: 978-1-4398-3088-8 (Paperback)

Visit the Taylor & Francis Web site at
http://www.taylorandfrancis.com

and the CRC Press Web site at
http://www.crcpress.com

This book is dedicated to Bill Cripps
— industrial chemist

Contents

Preface

This book is similar to previous Companion style books where each topic is covered in a single page outline format with enough detail to provide a good understanding of the subject.

This book emphasises the physics underlying chemistry, especially in the first chapter. I hope that by understanding what is happening from a physics point of view, the reader may better appreciate what is happening from the chemical perspective that is usually found in a traditional chemistry book.

I am indebted to Dr. Ray Sleet of the University of Technology, Sydney, an exemplary teacher who taught me that anything can be learned if it is broken down into manageable pieces with attainable goals. Many of the presentations in this book have their origin in Dr. Sleet's excellent undergraduate lectures.

I also thank Hilary Rowe for her persistence in getting this book into print and the editorial and production team at Taylor & Francis for their very professional and helpful approach to the whole publication process.

Tony Fischer-Cripps,
Killarney Heights, Australia

1. Structure of Matter

Summary

Energy levels Bohr H-like atom:

$$E_n = -\frac{m_e Z^2 q_e^4}{8\varepsilon_o^2 h^2 n^2}$$

Energy levels Schrödinger equation, Coulomb potential:

$$E_n = -\frac{Z^2 q_e^4 m}{(4\pi\varepsilon_o)^2 2\hbar^2 n^2}$$

Covalent bond: the co-sharing of valence electrons.

Ionic bond: the electrostatic attraction between ions formed after electrons are transferred from one atom to another.

Metallic bond: free electrons in the valence band reduce the energy of the system and so act to hold atoms together.

The electronegativity describes the relative ability of an atom, when it combines with another atom, to become "more negative" by more strongly capturing a shared electron or electron pair.

1.1 Atoms

Sixth century BC Thales of Miletos proposes that the basic element from which all things are made is water. 450 BC Empedocles teaches that all matter is composed of earth, air, water and fire. Around 400 BC, Greek philosophers (Leucippus, **Democritus**, Epicurus) proposed that if one could divide a piece of matter again and again, eventually a limit would be reached where no further subdivision could take place, this limiting amount of matter was called the **atom**. About 300 BC, **Aristotle** rejects the atomistic view and argues that matter is based upon the four basic elements of Empedocles but adds the qualities of coldness, hotness, dryness and moistness. Aristotle's considerable reputation ensured that his ideas became embodied in religious teaching for many hundreds of years.

In the fifteenth century, new advances in physics suggested that matter was made from particles, in agreement with the ancient Greek atomists. Robert **Boyle** taught that matter consists of different types of **elements** that were composed of atoms of the same type. Different elements could join together in fixed proportions to form **compounds**. Later, in 1803, John **Dalton** proposed the **atomic theory of matter** that was based on quantitative experimental evidence from the weighing of different elements in combination. He created a scale atomic mass for the different elements that were then known. Dalton's reference atom was the lightest element known, hydrogen, which was assigned an atomic mass of one. Other elements were given atomic masses according to how heavy they were compared to a hydrogen atom.

In 1807, Humphry Davy decomposed potash into sodium and potassium metals using electrolysis. In 1832, Michael **Faraday** discovered a quantitative connection between electricity and the separation of compounds into elements in electrolysis. These observations suggested that atoms themselves contain electric charge. Experiments by William Crookes demonstrated visible "cathode rays" that emanated from a negatively charged electrode (cathode) and travelled towards the positive electrode (anode) in an evacuated tube.

In 1872, **Mendeleev** arranged elements in increasing order of atomic mass and discovered that the properties of certain elements were repeated at regular intervals. When elements were ordered in columns with the atomic mass going across from left to right, and similar chemical properties going down, a **periodic table** was formed whereby, using the known elements at the time, Mendeleev was able to predict the properties of some as-yet undiscovered elements from gaps in the table.

1.2 Bohr Atom

In 1897, **Thomson** demonstrated that the rays observed to be emitted from the cathodes of vacuum tubes were in fact charged particles which he called **electrons**. Thomson proposed that atoms consisted of a positively charged sphere within which were embedded negatively charged electrons.

Rutherford found in 1911 that the electrons were actually located at some distance from a central positively charged **nucleus**. He proposed that electrons orbited the nucleus and the electrostatic attraction between the nucleus and the electron was balanced by the centrifugal force arising from the orbital motion. However, a major problem with this was that if this were the case, then the electrons would continuously radiate all their energy as **electromagnetic waves** and very quickly fall into the nucleus.

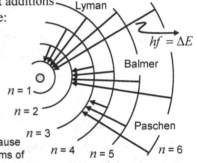

$$q_e = -1.6 \times 10^{-19} \, C$$

In 1913, **Bohr** postulated two important additions to Rutherford's theory of atomic structure:

1. Electrons of mass m_e can orbit the nucleus at radius r with velocity v in what are called **stationary states** in which no emission of radiation occurs and in which the **angular momentum** L is constrained to have values:

   $$L = m_e vr = \frac{nh}{2\pi}$$

   The 2π appears because L is expressed in terms of ω rather than f.

2. Electrons can make transitions from one state to another accompanied by the emission or absorption of a single **photon** of energy $E = hf$, this being the absorption and emission spectra observed experimentally.

As in the Rutherford atom, the centrifugal force is balanced by Coulomb attraction:

$$\frac{1}{4\pi\varepsilon_o} \frac{q_e^2}{r^2} = \frac{m_e v^2}{r}$$

with the addition that:

$$m_e vr = \frac{nh}{2\pi}$$

Mechanical model of hydrogen atom

By summing the **kinetic energy** (from the orbital velocity) and the **potential energy** from the electrostatic force, the **total energy** of an electron at a given energy level n is given by:

$$E_n = -\frac{m_e Z^2 q_e^4}{8\varepsilon_o^2 h^2 n^2}$$

Note: **atomic number** $Z = 1$ for the hydrogen atom where the energy of the ground state is −13.6 eV. The energy levels for each state n rise as Z^2. Thus, according to the Bohr model, the energy level of the innermost shell for multi-electron atoms can in principle be several thousand eV.

1.3 Energy Levels

The stationary states or energy levels allowed by the Bohr model are called **electron shells** or orbitals, and are labelled K, L, M, N, etc. with K corresponding to $n = 1$. The number n is called the **principal quantum number**. According to the Bohr model, the electron energy only depends on n, but experiments show that in multi-electron atoms, electron shells consist of sub-levels (evidenced by fine splitting of spectral lines). For example, the L shell $n = 2$ has two **sub-shells**, 2s and 2p.

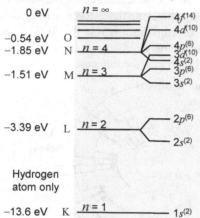

It is convenient to assign the energy at infinity as being 0 since as an electron moves closer to the nucleus, which is positively charged, its potential to do work is less and thus the energy levels for each shell shown are negative. In hydrogen, a single-electron atom, the energies for each shell are given by:

$$E = -\frac{13.6}{n^2} \quad \text{for hydrogen}$$

The electron-volt is a unit of energy.
1 eV = 1.602×10^{-19} J.

Sometimes the splitting of principal shells into sub-shells results in some overlap (e.g. 4s is lower in energy than 3d).

At each value of n the **angular momentum** can take on several distinct values. The number of values is described by the second quantum number l. The allowed values of l are 0, 1, ... $(n-1)$. Each value of l is indicated by a letter that indicates the sub-shell:

A third quantum number m describes the allowable changes in angle of the **angular momentum** vector in the presence of an electric field. It takes the values $-l$ to 0 to $+l$.

A fourth quantum number describes the **spin** of an electron where the spin can be either $-1/2$ or $+1/2$.

$l = 0$	s
$l = 1$	p
$l = 2$	d
$l = 3$	f
$l = 4$	g
$l = 5$	h

According to the **Pauli exclusion principle**, no electron in any one atom can have the same combination of quantum numbers. When all the electrons in an atom are in the lowest possible energy levels, the atom is said to be in its **ground state**. The outermost electrons in an atom are called the **valence electrons**.

For example, the 3d sub-shell can hold up to 10 electrons:

$$n = 3$$
thus: $l = 0, 1, 2 \ (s, p, d)$
and: $m = -2, -1, 0, 1, 2$

5 values of m times 2 for spin, thus 10 possible electrons

1.4 Schrödinger Equation

The total energy E of an electron in an atom is the sum of the potential and kinetic energies. Expressed in terms of **momentum**, p, and mass of electron m, this is stated:

$$E = \frac{p^2}{2m} + V$$

The value of the potential function may depend on both position and time. The form of $V(x,t)$ is different for different arrangements of atoms (e.g. a single isolated atom, an atom in a regular array of a crystal).

Thus: $$hf = \frac{p^2}{2m} + V(x,t)$$

since $E = hf$

Let $$p = -i\hbar \frac{\partial \Psi}{\partial x}$$

$$E = i\hbar \frac{\partial \Psi}{\partial t}$$

Ψ is a variable, the form and value of which provide information about the motion of a wave/particle.

Thus:

$$-\frac{\hbar^2}{2m}\frac{\partial^2 \Psi}{\partial x^2} + V(x,t)\Psi = i\hbar \frac{\partial \Psi}{\partial t}$$ **Schrödinger equation**

The solution to the Schrödinger wave equation is the **wave function** Ψ. If V is a function of x only, then the wave equation can be separated into time-independent and time-dependent equations that can be readily solved.

$$-\frac{\hbar^2}{2m}\frac{\partial^2 \psi}{\partial x^2} + V(\psi) = E\psi \qquad \phi(t) = e^{i\frac{E}{\hbar}t}$$

The resulting solutions of these equations, when multiplied together, give the **wave function**: $$\Psi(x,t) = \psi(x)\phi(t)$$

The wave function gives all the information about the motion of an electron in an atom. Ψ is a complex quantity, the magnitude of which $|\Psi|$ is interpreted as a **probability density function** which in turn can be used to determine the probability of an electron being at some position between x and Δx.

Quantum mechanics is concerned with determining the wave function (i.e. solving the Schrödinger equation) for particular potential energy functions such as those inside atoms. It is found that valid solutions to the time-independent wave equation occur only when the total energy is quantised. The solutions correspond to **stationary states**.

Solutions to the Schrödinger equation can be found for potential functions which are a function of both x and t. This enables time-dependent phenomena (e.g. the probability of transitions of electrons between energy levels in an atom) to be calculated and hence the intensity of spectral lines.

1.5 The Infinite Square Well

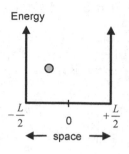

Consider an electron that is confined to be located in one-dimensional space $-L/2$ and $+L/2$ and where the **potential energy** of the electron is a constant (which can be conveniently set to zero). This is an example of an **infinite square well potential**. This potential is not usually found around electrons in an isolated atom, but does often represent that experienced by electrons in a solid and also in a chemical bond.

In the infinite square well potential, the electron cannot move more than a distance $x = \pm L/2$ from the centre position because it is constrained or bound by the infinite potential at the walls. In classical physics, the electron can have any value of total energy as it moves within this space.

In quantum physics, the allowed **stationary states** indicate the possible total energy of the electron. That is, the electron is moving as a particle whose probability of being in a particular position x is described by the standing wave patterns from the solutions to the **Schrödinger equation**. For the simple case of *zero* **potential energy** within the well, the energy levels are given by:

$$E = \frac{n^2 \pi^2 \hbar^2}{2L^2 m} \qquad n = 1,2,3,4\ldots, \ V(x) = 0$$

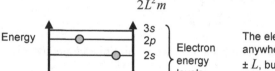

The electron can move anywhere within the confines $\pm L$, but can only have kinetic energies allowable by the stationary states.

It can be seen that the energy increases as n increases. That is, the electron has a greater **kinetic energy** if it exists in a **stationary state** with a larger n. The minimum allowable energy is greater than the minimum potential energy (in this case, 0). This is the **zero point energy**.

Note also that the kinetic energy of the electron, for a given value of n, decreases as the length of the well L increases. That is, if the electron is given more room to move, then its total energy is lowered (in this case, the total energy is kinetic, since the potential energy was set to be 0). "Available space" is effectively a mechanism for providing a reduction in total energy of the system. This is important when the **valence electrons** of different atoms come near each other during the formation of **chemical bonds**.

1.6 The Coulomb Potential

In the **infinite square well potential**, the potential energy of the electron inside the well was constant, and independent of position. In an isolated **single-electron atom**, a more realistic scenario is the case where electrons are bound by the **Coulomb potential**.

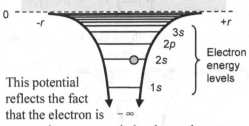

By convention, an electron is assigned a zero total energy when it is at rest, at an infinite distance from the nucleus. As the negatively charged electron moves towards the positively charged nucleus (opposite to the direction of the electric field), it can acquire kinetic energy and/or do work and so its electrical potential is reduced. That is, the potential energy becomes negative. This is the same convention in electric field theory, where it is usual to consider a *positive* test charge moving in the *same* direction of the electric field – whereupon potential energy is reduced.

This potential reflects the fact that the electron is attracted more strongly by the nucleus at shorter distances r and so the potential energy is no longer constant – i.e. the potential varies with distance from the nucleus. Mathematically, the Coulomb potential is:

$$V(r) = -\frac{Zq_e^2}{4\pi\varepsilon_o r}$$

where Z is the **atomic number**, the number of protons in the nucleus.

Note that in this potential, the energy (the potential energy) becomes more negative as r decreases. That is, electrons further away from the nucleus have a higher potential energy.

The total energy levels for an electron for this potential are given by the solution to the **Schrödinger equation** with $V(r)$ as above and are expressed:

$$E_n = -\frac{Z^2 q_e^4 m}{(4\pi\varepsilon_o)^2 2\hbar^2 n^2} \quad n = 1,2,3,4\ldots$$

This applies to an isolated **single electron atom** (e.g. hydrogen). Compare with Bohr model 1.2.

Here, the energy E becomes more negative as n decreases, and for a single electron ($Z = 1$), and at $n = 1$, we obtain the total energy $E = -13.6$ eV. The Coulomb potential, as drawn above, also reflects the greater range of movement available to the electron as r increases.

When an electron is bound by a potential, it is forced into having discrete allowable energies. That is, the allowable states $E < 0$ are **bound states**. Above this level, the electron is free and can have any energy. This is indicated by the grey band in the figure above. Note that in this potential, the energy depends on the principle quantum number n. As n increases, the energy becomes less negative and is therefore at a higher potential.

1.7 Covalent Bond

Consider a **hydrogen atom**.
Here we have one electron
orbiting the nucleus. In the
ground state, the electron is
in one of the available 1*s*
energy states.

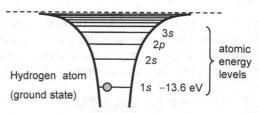

When a hydrogen atom comes into proximity with another hydrogen atom,
we might be tempted to draw the energy levels like this:

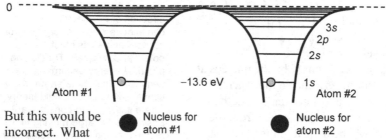

But this would be
incorrect. What
happens is that the electron in atom #2 is attracted by the positively charged
nucleus of atom #1 and vice versa. Overall, the nucleus from atom #1 is
attracted to both electrons. The nucleus from atom #2 is attracted to the same
two electrons. Therefore, the two nuclei behave as if they were bonded
together. The *co*-sharing of these *valence* electrons and the resulting
attraction of the two atoms is called a *covalent* bond.

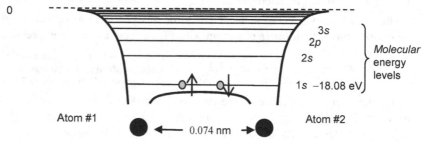

When this attraction occurs, the sides of the "well" are reduced so the
extent that the electrons are completely shared between the two atoms. The
electrons (by the **Pauli exclusion principle**) have to have different spins ↑↓.
Since this is energetically favourable (electrons have more space to move),
then hydrogen naturally forms the **molecule** H_2. When a covalent bond is
formed, energy is released (heat). To break the bond and separate the atoms,
energy $-(18.08-13.6) = -4.48$ eV has to be supplied.

1.8 Ionisation Energy

In order for an electron to make a transition from a lower state to a higher state in an atom, it must absorb energy (via collision with a photon of sufficient energy, by heat, etc). The energy required to move the outermost (highest potential energy) electron in an isolated atom from its ground state to infinity is called the **first ionisation energy** and is usually expressed in units of eV. The resulting positively charged atom is called an **ion**.

The energy required to move the next outermost bound electron from its **ground state** to infinity is called the second ionisation energy, and so on.

Sometimes ionisation energies are expressed as kJ per mole. Since 1 eV is 1.602×10^{-19} J, then 1 eV per atom is 96.49 kJ per mole.

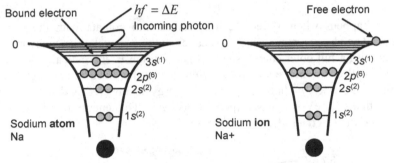

The incoming photon must have sufficient energy (in this example, $> +5.1$ eV for Na) to lift the electron to infinity distance from the nucleus for the atom to be ionised. If it does not, and if the incoming photon has sufficient energy to lift the electron to a higher energy level, then the atom may become "**excited**" rather than "**ionised**". If the incoming photon has insufficient energy to excite or ionise an atom, then it will just pass through or be scattered by the atom.

Important: We cannot just assign energies to these levels by calculation like we did with the hydrogen atom. The energies associated with each level in a multi-electron atom depend on the size of the atom, the atomic number, the degree of screening of outer electrons by the inner electrons, whether or not the atom is bonded with another, or exists in the gaseous phase, and most importantly, whether the level is filled or not. The ionisation energies are usually measured experimentally. The formula worked OK for hydrogen because we were dealing with just a single electron.

1.9 Electron Affinity

Consider a neutral chlorine atom. The first ionisation energy for Cl is 13 eV.

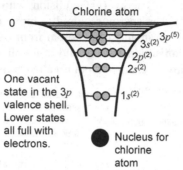

Chlorine atom

0

$3s^{(2)}$ $3p^{(5)}$
$2p^{(2)}$
$2s^{(2)}$

One vacant
state in the 3p $1s^{(2)}$
valence shell.
Lower states
all full with
electrons.   Nucleus for
                  chlorine
                  atom

What happens when a free electron comes into contact with the neutral atom? The electron may occupy a vacancy in the 3p level, thus endowing the atom with a full outer energy shell. This condition is energetically desirable and is called a **noble gas configuration**. When the electron is bought from infinity into the 3p shell, its potential energy is lowered and experiments show that −3.7 eV is released (perhaps as heat). This energy that is released is called the **electron affinity** of the atom.

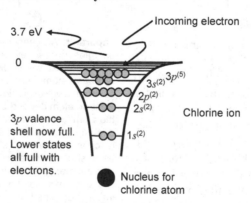

3.7 eV   Incoming electron

0

$3s^{(2)}$ $3p^{(5)}$
$2p^{(2)}$
$2s^{(2)}$ Chlorine ion

3p valence
shell now full. $1s^{(2)}$
Lower states
all full with
electrons.   Nucleus for
                  chlorine atom

The **electron affinity** is the energy released when an electron is bought from infinity into a neutral atom. Like **ionisation energy**, it is measured experimentally (although usually indirectly). The incoming electron has to overcome the repulsion of the electrons already there, but, in the case of elements like Cl, the attainment of a **noble gas configuration** is a sufficient payoff.

1.10 Ionic Bond

Ionic bonds usually form between elements that have unpaired valence electrons. Consider the reaction between sodium and chlorine.

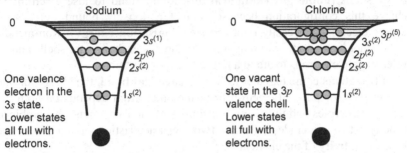

The ionisation energy for Na is +5.1 eV. The electron affinity for Cl is −3.7 eV. Thus, if the lone electron in the 3s band in Na can climb over the energy barrier (−5.1 − −3.7 = −1.4 eV), then this will create an Na⁺ sodium **ion** and a Cl⁻ chlorine ion, leaving each with a net charge q_e. The resulting electrostatic (Coulomb) attraction is called an **ionic bond**.

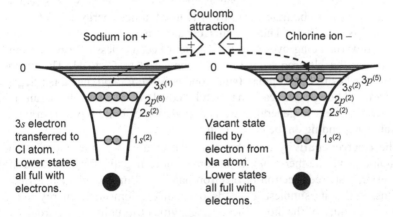

When the ions are formed, the attraction between them causes them to move towards each other and the electrical potential between them drops. The ions reach an **equilibrium distance** determined by the electrostatic attraction of their overall charge and the repulsion offered by their positively charged nuclei. Experiments show that the **bond energy** is −5.5 eV. This energy is lower than the −1.4 eV barrier and so an ionic bond between these two atoms is energetically favourable (since atomic systems tend settle to a state of minimum energy).

1.11 Electronegativity

It is energetically favourable for an atom to have completely filled outer energy shells (**noble gas configuration**). Metals tend to lose electrons to achieve this configuration because their few loosely bound outer shell valence electrons are easily removed (low **ionisation energy**). Non-metals, on the other hand, may have only a few vacancies in their outer shells and so prefer to gain electrons to attain a noble gas configuration.

When two atoms come together to form a **chemical bond**, the ability for one atom to gain, or pull, an electron from the other atom depends on both the ionisation energies and the electron affinities of the two atoms. A combined property that includes these two characteristics is called the **electronegativity** of the element.

Here's how it works in simplified terms. Hydrogen and chlorine can combine to form HCl by the formation of a covalent bond. An electron pair is shared between the H atom and the Cl atom. But is this sharing equal? The ionisation energy for H is +13.6 eV, and for Cl is +13 eV. The electron affinity of H is −0.75 eV and for Cl is −3.7 eV. Thus, for an electron to be transferred from Cl to H, a net energy of 13 − 0.75 = +12.25 eV is required. For an electron to be transferred from H to Cl, a net energy of 13.6 − 3.7 = +9.9 eV is required. Thus, in this covalent bond, the shared electron is biased towards being over near the Cl atom because less energy is required to transfer the electron from H to Cl compared to Cl to H. This unequal sharing makes the bond **polar** (since one end, the Cl end, has a net negative charge). Although we call the bond "covalent" it does have an ionic character as well. In general, there is a gradation of bond types from ionic to covalent depending on the nature of the atoms.

The **electronegativity** describes the relative ability for an atom, when it combines with another atom, to become "more negative" by more strongly capturing a shared electron or electron pair. It is measured by a variety of means, but at its simplest, depends on both the ionisation energy and the electron affinity of the atom. Electronegativities (no units) range from 4 (for Fl on the right-hand side of the periodic table) down to <1 (for elements on the left-hand side of the periodic table).

Two atoms with very different electronegativities are expected to form **ionic bonds**. Atoms with much the same electronegativities are expected to form **covalent bonds**.

An example is when an Na atom (with electronegativity 0.9) meets a Cl atom (electronegativity 3.0). The bond between them is expected to be predominantly ionic, with the chlorine atom becoming more negative. When H (with electronegativity 2.1) meets a Cl, the bond is expected to be predominantly covalent.

1.12 Metallic Bond

Consider what happens when we have two Li atoms close together. We might be first tempted to draw the potentials as:

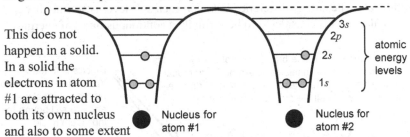

This does not happen in a solid. In a solid the electrons in atom #1 are attracted to both its own nucleus and also to some extent by the nucleus of atom #2. For example, the $1s$ electron orbiting the nucleus of atom #1 is also captured to some extent by the nucleus of atom #2. But, due to the **Pauli exclusion principle**, for each atom, we cannot have more than two electrons in the same energy level (i.e. the two $1s$ electrons for atom #1 as seen by atom #2 are no longer permitted to have energies at the $1s$ level because there are already two electrons from its own atom at that level).

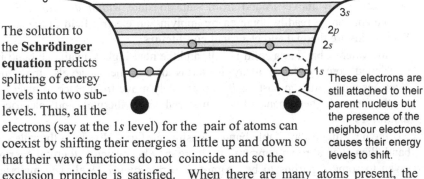

The solution to the **Schrödinger equation** predicts splitting of energy levels into two sub-levels. Thus, all the electrons (say at the $1s$ level) for the pair of atoms can coexist by shifting their energies a little up and down so that their wave functions do not coincide and so the exclusion principle is satisfied. When there are many atoms present, the splitting of many levels into fine gradations creates a **band** of energies and if the energy barrier between the atoms is low enough for the upper bands to carry across to meet those of neighbouring atoms, the electrons in these bands become free to migrate from atom to atom. The electrons behave as if they were in a **potential well**. Because of the large range of movement of these electrons (compared to if they were still attached to their parent nuclei), their kinetic energy is lower (see Section 1.5) and a lowering of energy equates to stability – that is, the free electrons in the valence band reduce the energy of the system and so act to hold the two atoms together. This is called a **metallic bond**.

1.13 Electronic Structure of Solids

In a solid, the interacting potentials of many millions of relatively closely spaced atoms causes atomic energy levels to split into a very large number of sub-levels. The energy difference between each sub-level is so fine that each molecular level is considered to be virtually a continuous **band** of energies.

In the diagram here, the broadening of the $2p$ level is such that electrons in this band are no longer local to a particular atom. These electrons are effectively shared between all the atomic nuclei present.

Electrons in these bands are constrained by the potential well and are still bound to individual nuclei.

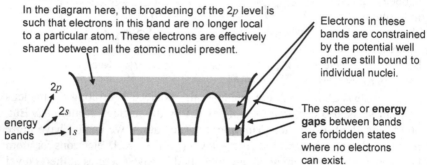

The spaces or **energy gaps** between bands are forbidden states where no electrons can exist.

If the highest energy band that contains electrons in the ground state (the **valence band**) in a solid is not completely full, then electrons within that band can easily move around from state to state *within the band*. Such movement can be readily obtained by applying an electric field to the solid. Such solids are thermal and electrical **conductors**.

If the valence band in a solid is full, and the next highest available band is positioned some distance away in terms of its energy levels, then the electrons within the topmost band cannot easily move from place to place or to the next highest band. Such materials are thermal and electrical **insulators**.

If the next highest available band is positioned fairly closely to the valence band, then even at room temperature, there may be sufficient thermal energy given to some electrons to be promoted to this higher level. The material becomes conducting and is a **semiconductor**. The band containing the conducting electrons is called the **conduction band**. In a conductor, the valence band is the conduction band. In a semiconductor, the conduction band (at 0K) is separated from the valence band (defined at 0K) by an **energy gap**.

Atoms in a solid generally form molecules which either arrange themselves in a regular pattern (**crystalline solids**) or the molecules do not repeat themselves in an orderly way (**amorphous solids**). When a solid forms, atoms are pushed and pulled around and settle into place when the interaction electron potentials reach a minimum level.

2. Chemical Components

Summary Mass number – total number of protons and neutrons in the nucleus

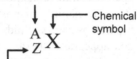

Atomic number – number of protons in the nucleus

One atomic mass unit (amu) is 1/12th the mass of a carbon 12 atom. 1 amu = 1.6602×10^{-27} kg.

Avogadro's number: 6.022×10^{23} = 1 mole.

Atoms that lose or gain electrons are called ions:

- Cations (+) (electrons lost)
- Anions (–) (electrons gained)

A molecule is the smallest collection of atoms that is electrically neutral and can exist as a separate identifiable unit.

The molecular weight of a substance is the sum of the atomic weights of its constituents.

Stoichiometry is the process of accounting for the masses of atoms, molecules and compounds in chemical reactions.

2.1 Matter

Elements: cannot be reduced to simpler units by physical processes or chemical reactions. (Examples are oxygen, iron, carbon.) Generally, elements are substances which have only one kind of atom (as having a certain number of protons on the nucleus). There are just over 100 elements known, with about 95% of these occurring naturally, the remainder being created synthetically by nuclear processes.

Elements can be broadly divided into non-metals, semi-metals and metals.

Mixtures: can be separated into constituent parts by physical means. Mixtures are a product of mechanical or physical processes, such as metal alloys, suspensions, dispersions and colloids. **Homogenous** mixtures have uniform composition and appearance (examples are air, salty water) over a molecular length scale while **heterogeneous** mixtures have physically distinct regions. (Examples are fruit cake, concrete, fruit juice with pulp.)

Compounds: combinations of two or more elements in definite proportions where the composition of the compound is uniform throughout. Compounds have a chemical structure, the atoms being held in place by chemical bonds. Compounds keep their chemical identity when altered physically, and can only be reduced to their constitutive elements by chemical reactions. (Examples are water, salt, sulphuric acid.)

2.2 Atomic Weight

Except in the case of nuclear transformations, we can usually treat atoms as being much like the Bohr atom. Atoms are characterised by an **atomic number**, a **chemical symbol**, and the **mass number**.

$$m_e = 9.1096 \times 10^{-31} \text{ kg}$$

Mass number – total number of protons and neutrons in the nucleus

$$m_p = 1.6726 \times 10^{-27} \text{ kg}$$

$$m_N = 1.6749 \times 10^{-27} \text{ kg}$$

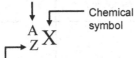

$$^A_Z X$$

Chemical symbol

Atomic number - number of protons in the nucleus

The mass number is approximately equal to the atomic weight of the element. The **atomic weight** of an element is the mass (in relative terms) of an "average" atom of an element. The more precise term **atomic mass** is used to denote the absolute mass of a particular atom (in kg).

Since the **mass of an electron** m_e is very small nearly all of the mass of an atom is contributed by the **protons** and **neutrons** in the nucleus. Protons and neutrons have very nearly the same mass, so the number of neutrons in an atom is given by $Z - A$. However, it is found in nature that many atoms of the same element with atomic number Z have different mass numbers A because of having a different number of neutrons in the nucleus. Each type of atom A of an element is called a **nuclide**, or **isotope**.

By international agreement, one **atomic mass unit (amu)** is 1/12th the mass of a single carbon 12 atom. The atomic weight (or **relative atomic mass**) of an element is the ratio of the average mass per atom of the element to 1/12 of the mass of an atom of ^{12}C.

Note: The atomic weight of an element is not found from adding the masses of protons, neutrons and electrons. When an atom is formed, some of the mass is used as nuclear binding energy via the **Einstein** relationship $E = mc^2$.

The **carbon 12** isotope has 6 protons and 6 neutrons in the nucleus. The **carbon 13** atom has 6 protons and 7 neutrons in the nucleus. Both isotopes have 6 electrons. Experiments show that carbon in nature consists of 1.11% ^{13}C and 98.89% ^{12}C. The atomic weight of carbon is determined to be 12.011.

In chemistry, much of the arithmetic of chemical reactions involves the atomic weights for the atoms that take part. However, the actual nature of the chemical reactions is due to the number and arrangement of the outer-shell **valence electrons** in the atoms rather than the heavy nuclei. "Valence" is a word that means "power" and in some sense, it is the number and arrangement of valence electrons that give an atom "combining power" to form **chemical bonds** with other atoms.

2.3 Ions

Atoms that lose or gain electrons are called **ions**. When an atom loses one or more electrons, it is called a **cation**. When an atom gains one or more electrons, it is called an **anion**.

Examples of ions:

Atom	Z (atomic number)	No. electrons gained or lost	No. electrons remaining	Formula of ion
H	1	-1	0	H^+
Na	11	-1	10	Na^+
Ca	20	-2	18	Ca^{2+}
F	9	+1	10	F^-
S	16	+2	18	S^{2-}

When an ion is formed, the atom is no longer electrically neutral. The formula of the ion signifies the net electronic charge. Atoms may lose electrons, especially unpaired outer shell or valence electrons, by physical removal, absorption of **ionising radiation**, the electron being taken by another type of atom which has a stronger attraction for it, and so on.

Metallic elements tend to lose one or more electrons when they combine with other elements and so form cations. Non-metallic elements tend to gain electrons to form anions. When naming compounds, it is usual to list the more metallic element first. In **binary ionic compounds**, we list the metal cation first followed by the non-metal anion with an "ide" suffix. For covalent compounds, the more metallic element is listed first.

Sodium chloride
Carbon monoxide
Carbon dioxide   } Where there is more than one
Nitrogen oxide   possible compound, prefixes are
                        used to distinguish them.

2.4 Molecules

Most elements bond with others to form **molecules**. A molecule is the smallest collection of atoms, that is electrically neutral, and can exist as a separate identifiable unit. The concept was first proposed by **Avogadro** in 1811, who, on the basis of observations made by Dalton and Gay-Lussac proposed that atoms combine to form molecules, *and* at the same temperature and pressure, equal volumes of *all* gases contain the same number of molecules. **Avogadro's law** provides the justification for a method of determining relative molecular weights — because there was now a way to obtain equal numbers of different molecules by using different gases all at the same temperature, pressure and volume.

A molecule is most conveniently described in terms of a **molecular formula**. H_2O A water molecule consists of two H atoms and one O atom.

In some cases, such as in a crystalline solid, the concept of a molecule is not appropriate because the atoms which make up the substance are arranged in a regular repeating pattern. In this case, we speak of the **empirical formula**.

$NaCl$ A sodium chloride crystal consists of equal numbers of sodium and chlorine atoms arranged in a regular array or crystal lattice.

The **molecular weight** (or **relative molar mass**, or **relative molecular mass**) M_r of a substance is the sum of the atomic weights of its constituents.

If we had a certain mass of an element (or a molecule), say 12 grams of ^{12}C, how many atoms would this be? This number is called **Avogadro's number** N_A and has the value 6.0220943×10^{23}. N_A is found by experimental methods.

By International agreement, one **atomic mass unit (amu)** is 1/12th the mass of a carbon 12 atom.

"But!", you might say, "How can you have 6.022... atoms if atoms are indivisible? Shouldn't this be a whole number?" Look at the 10^{23}. If you write Avogadro's number out in full, it *is* a whole number.

12 g of ^{12}C = 6.0220943×10^{23} atoms; therefore the mass of one ^{12}C atom is:

$$\text{mass} = \frac{0.012}{6.02 \times 10^{23}}\,\text{kg}$$

$$= 1.99 \times 10^{-23}\,\text{kg}$$

It is convenient when working with quantities usually involved in chemical reactions to round down Avogadro's number to 6.02×10^{23}.

2.5 Mole

Chemical reactions usually involve large numbers of atoms and molecules. It is convenient to have a unit of measure that relates weight (i.e. mass in grams or kilograms, etc) of bulk chemicals with atomic or molecular weight. By definition, this unit of measure is called a **mole**, and it is the Avogadro number of atoms or molecules.

12 grams of ^{12}C contains 6.02×10^{23} atoms and is called 1 mole of ^{12}C

18 grams of H_2O also contains 6.02×10^{23} molecules and is called 1 mole of H_2O

The word "mole" is a shorthand way of saying "**Avogadro's number**". That is, it is easier to say "Consider one mole of sodium atoms" rather than "Consider one Avogadro's number of sodium atoms" or "Consider 6.02×10^{23} of sodium atoms". To find out the mass in grams of a mole of some atoms, we simply look up the
atomic weight. ⎯⎯⎯⎯⎯⎯⎯⎯⎯▶ because the atomic weight (in
To find out the mass in grams of a   amu) is expressed relative to
mole of some molecules, we simply   the ^{12}C atom, and thanks to
use the molecular weight.   Avogadro, we know that 12 g of
                                                                 ^{12}C has N_A atoms.

Moles, why bother? It's convenient. One mole of *anything* is 6.02×10^{23} of the "anythings". When we are talking about atoms or molecules, we know that one mole has a mass equal to the atomic weight of the element, or the molecule because that's the way Avogadro's number was determined. It's the link between the *relative* atomic mass and the *actual* mass in grams.

Units of chemical accounting:

- One **atomic mass unit** (amu) is 1/12th the mass of a carbon 12 atom by international agreement. 1 amu is $1.66020943 \times 10^{-27}$ kg.

- The **atomic weight** (or **relative atomic mass**) of an element is the ratio of the average mass per atom of the element to 1/12 of the mass of an atom of ^{12}C.

- The total sum of the atomic weights for a molecule of substance is called the **molecular weight** or **relative molar mass** M_r.

- The molecular/atomic weight expressed in grams contains one mole of molecules/atoms. This is Avogadro's number 6.0220943×10^{23}.

- The **molar mass** M of an element is the atomic weight of the element expressed in g/mol. The molar mass of a compound is the sum of the atomic weights of the constituent elements expressed in g/mol.

2.6 Compounds

Compounds consist of two or more different elements which are chemically combined in a definite fixed ratio (as distinct from mixtures, which can have any proportions of ingredients). There are generally two types of compounds:

Molecular compounds are composed of atoms held together by chemical bonds in fixed proportion and are electrically neutral overall. For example, the water molecule H_2O is a molecular compound and the **molecular formula** shows the proportions of the atomic species within it.

Name	Molecular formula	
Methane	CH_4	There is a distinction to be made between the **molecular formula** and the **empirical formula** of a compound. The empirical formula gives the ratio of component atoms in the lowest possible numerical terms. The empirical formula shows the ratio of component atoms as found to exist in practice. In many cases, the formulae are the same, but this cannot be always assumed. For example, the empirical formula for benzene is CH but the molecular formula is C_6H_6.
Carbon dioxide	CO_2	
Ammonia	NH_3	
Acetylene	C_2H_2	
Ethylene	C_2H_4	
Ethane	C_2H_6	
Water	H_2O	
Benzene	C_6H_6	
Ethanol	C_2H_5OH	
Naphthalene	$C_{10}H_8$	
Aspirin	$C_9H_8O_4$	

Ionic compounds are composed of charged atoms, or a charged group of atoms, of opposite sign that are held together by **ionic bonds**. When an ionic compound is formed from the joining of a charged group of atoms, it is said to be a **complex ion** (or a polyatomic ion, or a molecular ion). The ion with the positive charge in the compound is the **cation**. The ion with the negative charge is the **anion**. The ionic formula gives the ratio of anions and cations. The ionic formula is electrically neutral overall.

Name	Ions formed	Formula
Magnesium chloride	Mg^{2+} Cl^-	$MgCl_2$
Silver sulphate	Ag^+ SO_4^{2-}	Ag_2SO_4
Ammonium sulphate	NH_4^+ SO_4^{2-}	$(NH_4)_2SO_4$
Chromium (III) hydroxide	Cr^{3+} OH^-	$Cr(OH)_3$
Sodium chloride	Na^+ Cl^-	$NaCl$

2.7 Cations

+1

Name	Formula
Hydrogen	H^+
Lithium	Li^+
Sodium	Na^+
Potassium	K^+
Rubidium	Rb^+
Caesium	Cs^+
Silver	Ag^+
Copper	Cu^+
Mercury (I)	Hg_2^{2+}
Ammonium	NH_4^+

+2

Name	Formula
Beryllium	Be^{2+}
Magnesium	Mg^{2+}
Calcium	Ca^{2+}
Strontium	Sr^{2+}
Barium	Ba^{2+}
Lead	Pb^{2+}
Zinc	Zn^{2+}
Cadmium	Cd^{2+}
Nickel	Ni^{2+}
Manganese	Mn^{2+}
Tin (II)	Sn^{2+}
Iron (II)	Fe^{2+}
Mercury (II)	Hg^{2+}
Cobalt (II)	Co^{2+}
Chromium (II)	Cr^{2+}
Copper (II)	Cu^{2+}

+3 , +4

Name	Formula
Aluminium	Al^{3+}
Bismuth	Bi^{3+}
Iron (III)	Fe^{3+}
Cobalt (III)	Co^{3+}
Chromium (III)	Cr^{3+}
Tin (IV)	Sn^{4+}

In the naming of chemical compounds, it is customary to write the name of the most metallic element first, followed by the more non-metallic one. The second element or ion is given the *ide* suffix (e.g. sodium chloride).

Prefixes
1 – mono
2 – di
3 – tri
4 – tetra
5 – penta
6 – hexa

2.8 Anions

−1

Name	Formula
Fluoride	F^-
Chloride	Cl^-
Bromide	Br^-
Iodide	I^-
Hydroxide	OH^-
Nitrite	NO_2^-
Nitrate	NO_3^-
Chlorate	ClO_3^-
Perchlorate	ClO_4^-
Cyanide	CN^-
Permanganate	MnO_4^-
Thiocyanate	NCS^-
Bicarbonate	HCO_3^-
Bisulphate	HSO_4^-
Dihydrogen phosphate	$H_2PO_4^-$
Acetate	$CH_3CO_2^-$

−2

Name	Formula
Oxide	O^{2-}
Peroxide	O_2^{2-}
Sulphide	S^{2-}
Sulphate	SO_4^{2-}
Sulphite	SO_3^{2-}
Thiosulphate	$S_2O_3^{2-}$
Carbonate	CO_3^{2-}
Oxalate	$C_2O_4^{2-}$
Chromate	CrO_4^{2-}
Dichromate	$Cr_2O_7^{2-}$
Monohydrogen phosphate	HPO_4^{2-}

−3 , −4

Name	Formula
Nitride	N^{3-}
Phosphate	PO_4^{3-}
Hexacyanoferrate (III)	$Fe(CN)_6^{3-}$
Hexacyanoferrate (II)	$Fe(CN)_6^{4-}$

For polyatomic anions, the assignment of a suffix depends on the **oxidation number** (see Section 4.8) of the central non-metal atom.

Suffixes	
ide	
ite	ous
ate	ic

In some cases, the oxidation number of the metal is stated in roman numerals to distinguish compounds which would otherwise have the same name (e.g. iron (II) chloride and iron (III) chloride).

2.9 Chemical Equation

When atoms or molecules (the **reactants**) combine with other atoms or molecules to form compounds (the **products**), a **chemical reaction** is said to take place. Chemical reactions involve the breaking and formation of **chemical bonds**.

$$\text{Reactants} \rightarrow \text{Products}$$
This is the
forward reaction

$$\text{Products} \rightarrow \text{Reactants}$$
This is the
reverse reaction

Often one can tell if a chemical reaction has occurred by observation of products that are considerably different in appearance than the reactants. Examples are the formation of a **precipitate**, evolution of gas, change of colour, production of heat and so on. Often, chemical reactions proceed spontaneously, other times they have to be forced to proceed by the input of energy or the lowering of energy barriers.

Chemical equations are written using chemical formulae where the number of atoms or molecules is written before the formula:

$$2CO + O_2 \rightarrow 2CO_2$$

This equation says that two molecules of carbon monoxide react with one molecule of oxygen to product two molecules of carbon dioxide. This is a **balanced equation**, because the number of atoms in each reactant equals the number of atoms in each product. That is, we have two carbon atoms on the left-hand side and two carbon atoms on the right-hand side, and two oxygen atoms on the left side and two on the right side. We can also say that two **moles** of carbon monoxide combine with one mole of oxygen to form two moles of carbon dioxide. Since the mass of a mole of an atom or a molecule is the atomic or molecular weight expressed in grams, a balanced equation is balanced in terms of both number of atoms and mass on both sides of the equation. The net electric charge must also be the same on both sides of the equation.

Usually, reactions proceed until there is **chemical equilibrium**. Dynamic chemical equilibrium occurs when the rate of the forward reaction becomes equal to the rate of the reverse reaction:

$$\text{Reactants} \leftrightarrow \text{Products}$$
Dynamic
equilibrium

By rate, we mean the rate of formation of products in units of mol/s.

2.10 Stoichiometry

Stoichiometry is the process of accounting for the masses of atoms, molecules and compounds in chemical reactions. The basic component of such accounting is the arrangement of chemical formulae in chemical equations. Stoichiometric calculations are performed in a logical sequence:

1. Write a balanced equation for the reaction.
2. Convert known masses of reactants and/or products into moles.
3. Calculate the number of moles of the reactants and/or products whose masses are unknown.
4. Calculate the masses of the unknown reactants and/or products.

Balancing an equation involves both a balance of atoms and also electric charge. This is particularly important in the case of ionic equations where spectator ions may be present and do not take part in the reaction. Some examples of unbalanced equations (with spectator ions) and the corresponding **net ionic equations** are:

$$Pb(NO_3)_2\,(aq) + KCl(aq) \rightarrow PbCl_2\,(s) + K^+ + NO_3^{3-}$$
$$Pb^{2+}\,(aq) + 2Cl^-\,(aq) \rightarrow PbCl_2\,(s)$$

$$Ni(OH)_2\,(s) + HNO_3(aq) \rightarrow NiNO_3\,(aq) + H_2O$$
$$Ni(OH)_2\,(s) + 2H^+ \rightarrow Ni^{2+} + 2H_2O$$

$$NaCO_3(aq) + HNO_3(aq) \rightarrow NaNO_3(aq) + H_2O + CO_2$$
$$CO_3^{2-} + 2H^+ \rightarrow 2H_2O + CO_2$$

Practical difficulties can arise when attempting to determine masses of reactants and products. For example, some reactants may be present in excess (more present than combine with one or other reactants). More than one reaction may take place at the same time, or a reaction may proceed in a series of intermediate steps. Products (such as gases) may escape and alter the equilibrium state of the reaction. The nature of the reactants may prevent mixing of the reactants and so there is incomplete yield of product.

2.11 Example

Acetylene gas is used together with oxygen in an "oxy" welding set. It is prepared from a reaction of calcium carbide and water:

$$CaC_2 + H_2O \rightarrow Ca(OH)_2 + C_2H_2 \quad \text{Unbalanced eqn.}$$

Balance this equation and determine the mass of acetylene that is produced from the reaction of 130 g CaC_2 and 100 g water.

$$CaC_2 + 2H_2O \rightarrow Ca(OH)_2 + C_2H_2 \quad \text{Balanced eqn.}$$

$$CaC_2 = 130\,g$$

$$= \frac{130}{40.1 + 2(12)}$$

$$= 2.03\,\text{moles}$$

$$H_2O = 100\,g$$

$$= \frac{100}{2 + 16}$$

$$= 5.56\,\text{moles}$$

But according to the balanced equation, for each mole of CaC_2, two moles of H_2O are required. Therefore, $2(2.03) = 4.06$ moles of H_2O required, and so the mass of C_2H_2 formed would be:

$$C_2H_2 = 2.03\,\text{moles}$$

$$2.03 = \frac{m}{2(12) + 2(1)}$$

$$= 52.8\,g$$

3. The Periodic Table

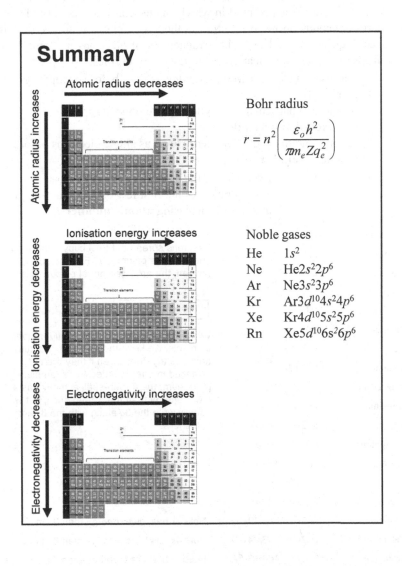

Summary

Atomic radius increases

Atomic radius decreases

Ionisation energy decreases

Ionisation energy increases

Electronegativity decreases

Electronegativity increases

Bohr radius

$$r = n^2 \left(\frac{\varepsilon_o h^2}{\pi m_e Z q_e^2} \right)$$

Noble gases

He	$1s^2$
Ne	$He2s^22p^6$
Ar	$Ne3s^23p^6$
Kr	$Ar3d^{10}4s^24p^6$
Xe	$Kr4d^{10}5s^25p^6$
Rn	$Xe5d^{10}6s^26p^6$

3.1 Electron Configuration

In a **multi-electron atom**, the electrons do not find themselves in a **Coulomb potential** of the form $1/r$ as in the **hydrogen atom** due to the **screening effect** of inner-shell electrons on the outer-shell electrons. In a multi-electron atom, the potential in which an electron finds itself (and hence its energy) depends on not only n (as in the single-electron atom), but also the second quantum number l. The arrangement of electrons in atoms in the **ground state** can be complicated, but follows certain rules:

1. Electrons occupy the lower energy levels first, before occupying higher levels.
2. Electrons at a particular energy level have opposite spin (**Pauli exclusion principle**).
3. Electrons tend to occupy an energy level as single electrons before pairing up (**Hund's rule**).

Hydrogen	$1s^1$	Examples of the ground-state electron
Helium	$1s^2$	configuration of the first few elements arranged
Lithium	$1s^2 2s^1$	in order of increasing **atomic number** Z

Beryllium $\quad 1s^2 2s^2$

Boron $\qquad 1s^2 2s^2 2p^1$

Carbon $\qquad 1s^2 2s^2 2p^2$

Nitrogen $\quad 1s^2 2s^2 2p^3$

Oxygen $\qquad 1s^2 2s^2 2p^4$

Fluorine $\quad 1s^2 2s^2 2p^5$

Neon $\qquad 1s^2 2s^2 2p^6$

Sodium $\qquad 1s^2 2s^2 2p^6 3s^1$

Magnesium $\quad 1s^2 2s^2 2p^6 3s^2$

Aluminium $\quad 1s^2 2s^2 2p^6 3s^2 3p^1$

Silicon $\qquad 1s^2 2s^2 2p^6 3s^2 3p^2$

Phosphorus $\quad 1s^2 2s^2 2p^6 3s^2 3p^3$

Sulphur $\qquad 1s^2 2s^2 2p^6 3s^2 3p^4$

Chlorine $\quad 1s^2 2s^2 2p^6 3s^2 3p^5$

Argon $\qquad 1s^2 2s^2 2p^6 3s^2 3p^6$

Potassium $\quad 1s^2 2s^2 2p^6 3s^2 3p^6 4s^1$

Calcium $\qquad 1s^2 2s^2 2p^6 3s^2 3p^6 4s^2$

Scandium $\quad 1s^2 2s^2 2p^6 3s^2 3p^6 3d^1 4s^2$

Electrons spread out to occupy unfilled sub-shells before pairing up. For example, in oxygen, the arrangement of electrons is:

When we write the electron configuration of elements, we keep the principal quantum numbers together even if they are not arranged in energy order. For example, in potassium, the $4s$ sub-shell actually has a lower energy than the $3d$ sub-shell, but we usually write the $3d$ along with the $3s$ and $3p$ sub-shells.

Note $4s$ has lower energy than $3d$ in K.

Note $4s$ has lower energy than $3d$ in Ca.

Keep sub-shells together even if no longer in increasing energy order.

3.2 Periodic Law

In the nineteenth century, it was noticed that if elements were arranged in order of increasing **atomic weight**, certain properties of the elements tended to occur at periodic intervals. This is a striking observation. It is now known that **periodicity** occurs when the elements are arranged in increasing **atomic number** Z. Consider the first few elements and the property of reactivity:

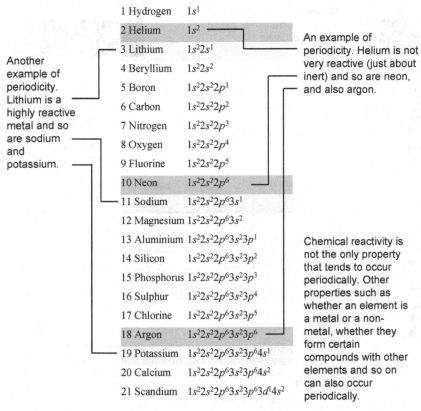

1 Hydrogen	$1s^1$	
2 Helium	$1s^2$	
3 Lithium	$1s^2 2s^1$	
4 Beryllium	$1s^2 2s^2$	
5 Boron	$1s^2 2s^2 2p^1$	
6 Carbon	$1s^2 2s^2 2p^2$	
7 Nitrogen	$1s^2 2s^2 2p^3$	
8 Oxygen	$1s^2 2s^2 2p^4$	
9 Fluorine	$1s^2 2s^2 2p^5$	
10 Neon	$1s^2 2s^2 2p^6$	
11 Sodium	$1s^2 2s^2 2p^6 3s^1$	
12 Magnesium	$1s^2 2s^2 2p^6 3s^2$	
13 Aluminium	$1s^2 2s^2 2p^6 3s^2 3p^1$	
14 Silicon	$1s^2 2s^2 2p^6 3s^2 3p^2$	
15 Phosphorus	$1s^2 2s^2 2p^6 3s^2 3p^3$	
16 Sulphur	$1s^2 2s^2 2p^6 3s^2 3p^4$	
17 Chlorine	$1s^2 2s^2 2p^6 3s^2 3p^5$	
18 Argon	$1s^2 2s^2 2p^6 3s^2 3p^6$	
19 Potassium	$1s^2 2s^2 2p^6 3s^2 3p^6 4s^1$	
20 Calcium	$1s^2 2s^2 2p^6 3s^2 3p^6 4s^2$	
21 Scandium	$1s^2 2s^2 2p^6 3s^2 3p^6 3d^1 4s^2$	

Another example of periodicity. Lithium is a highly reactive metal and so are sodium and potassium.

An example of periodicity. Helium is not very reactive (just about inert) and so are neon, and also argon.

Chemical reactivity is not the only property that tends to occur periodically. Other properties such as whether an element is a metal or a non-metal, whether they form certain compounds with other elements and so on can also occur periodically.

Initially it seems that the chemical reactivity of elements repeats in intervals of eight in atomic number Z. However, when the elements are arranged more completely we see that the periodicity actually depends upon the arrangement of outer-shell **valence electrons** – a not surprising observation in hindsight because it is the valence electrons that are involved in the formation of **chemical bonds**. For example, neon (10), argon (18), krypton (36), xenon (54) and radon (86) all have eight electrons in the outer shell and are inert. Na and K have one valence electron and easily form positive ions.

3.3 Periodic Table

Periodicity in chemical (and sometimes physical) properties can more readily be appreciated when elements are arranged in a table. Vertical columns in the table (**groups**) show elements with similar properties. Each horizontal row is called a **period**. The table also indicates (approximately) the way in which electron shells are occupied.

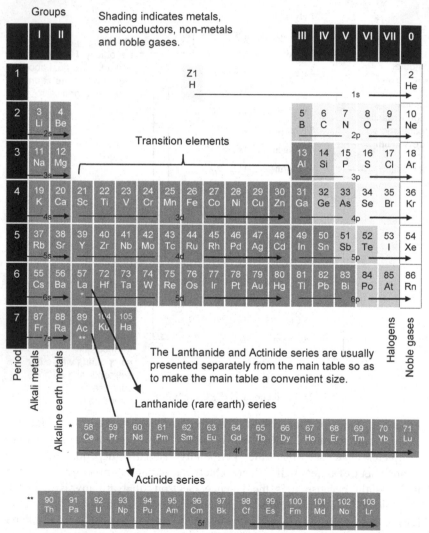

3.4 Groups

Hydrogen is placed in a separate position in the periodic table.

Hydrogen has one electron, which in the ground state is in the $1s$ orbital. This unpaired electron makes hydrogen a reactive element that is able to form compounds with many other elements, including itself, to form a gas H_2.

^1_1H Protium 99.9% abundance

^2_1H Deuterium (heavy hydrogen)

^3_1H Tritium (radioactive)

Group I elements are the **alkali metals**. Having one weakly bound outer valence electron, they are highly reactive, and metallic in character. Elements in this group usually form M^+ ions which are water soluble.

Group II elements are the **alkaline earth metals**. Elements in this group usually form M^{2+} ions and are generally insoluble, occurring naturally as silicates, carbonates, sulphates and phosphates.

The **transition elements** generally occupy the positions between Groups II and III. All these elements are metals since their outermost shells contain only a few electrons. However, unlike alkali metals, transition metals are hard, brittle and have a high melting point (with the exception of mercury).

Group III elements are also considered metals (although boron has only semi-metallic properties). With the exception of boron, they form M^{3+} ions and are relatively soft.

Group IV elements range from non-metal, semi-metals (or semiconductors) to metals down the group with increasing size of atom (and hence screening effect of electrons). All have four electrons in their outer shell. **Carbon** is responsible for the formation of hydrocarbons and derivatives, the basis of life on Earth. **Silicon**, unlike carbon, tends to form bonds with oxygen and is the basis for most of the minerals of the Earth.

Group V elements range from non-metallic to metallic down the group with increasing atomic size and five valence electrons.

Group VI elements show little metallic character due to the increasing ionisation potentials as we go across the periodic table. Increasing atomic size going down the group confers some metallic properties to Se, Te and Po.

Group VII elements are referred to as the halogens. They all have a high tendency to complete their electron shells by forming salts. These elements are largely non-metallic, and mostly reactive.

Group 0 elements are the **noble gases**, so called because of their inertness, although they are able to form compounds with oxygen and fluorine under certain conditions. Argon was discovered in 1894.

3.5 Energy Levels

The ordering of the outer electron shells in terms of energy is the basis for interpretation of the **periodic table**. Calculations show that the ordering of the energy shells proceeds in the following sequence from lowest (more negative) to highest (less negative potential):

$1s, 2s, 2p, 3s, 3p, 4s, 3d, 4p, 5s, 4d, 5p, 6s, 4f, 5d, 6p, 7s, 5f, 6d$

From $1s$ to $3p$, it is easy to understand because the electron shells are at an increasing radial distance from the nucleus (n quantum number dominates the energy distribution just as in the one-electron atom). The n quantum number determines the distance the electrons are from the nucleus. The l quantum number determines the shape of the electron shell. When $l = 0$, we have an s sub-shell which is spherical. When $l = 1$, we have a p sub-shell which is lobed. $l = 2$ gives d sub-shells which are four-lobed geometries. The significance of this is that in the fourth period, in K and Ca, the $4s$ sub-shell has a lower energy than the $3d$ sub shell. In

$l = 0$	s
$l = 1$	p
$l = 2$	d
$l = 3$	f
$l = 4$	g
$l = 5$	h

these elements, electrons prefer to be in $4s$ rather than $3d$ in the ground state even though the radius of the $4s$ shell is larger than the $3d$ shell.

For the first row of the **transition elements**, the $3d$ levels become occupied only after the $4s$ levels are filled (except for Cr). In these elements, the electrons in the $4s$ shell, being further from the nucleus (but having lower energy than the $3d$ electrons) shield those in the inner $3d$ shell and so the chemical properties for these elements are all very similar since in each case, the valence electrons are the $4s$ *outer* electrons while it is the number of inner $3d$ electrons that is changing. It is the outer **valence electrons** (with the highest principal quantum number) that interact with other atoms to form chemical bonds.

The ordering of shells given above is not the ordering of *all* the energy levels *within* an atom, only the *energy* ordering of the *outer* shells. For example, in K, the $4s$ shell is at a lower energy than the $3d$ shell ($3s^2 3p^6 4s^1$). By the time we get to Zn, the $3d$ shell is at a lower energy than the $4s$ shell ($3d^{10} 4s^2$). There is no one sequence of energy shells that applies to all elements.

This is a consequence of the increasing dominance of the l quantum number in the d sub-shell in determining the energy for a sub-shell. The l quantum number is connected with the angular momentum of the electrons and so, much like the case where the angular momentum for a rotating wheel is greater if the mass is concentrated at the outer edge compared to the case when the mass is evenly distributed, the concentration of mass of electrons in d-shaped shells results in a greater kinetic energy component to the total energy compared to spherical s shells.

3.6 Noble Gases

The **noble gas** elements are extremely stable and generally do not form compounds with any other elements. The notable feature of noble gas elements is the interesting property of having full outer electron energy shells. Consider the second period elements. As we go from left to right, the number of electrons in the energy level with principal quantum number $n = 2$ is increasing. At neon, we have eight valance electrons: $2s^2\ 2p^6$. A group of eight (**octet**) valence electrons that completely fills an energy shell is very energetically stable and is called the **noble gas configuration**.

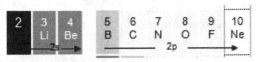

An atom with one more valence electron makes the element sodium, a highly reactive metal. Sodium is highly reactive because its single $3s^1$ valence electron is easily removed so that the sodium **cation** has a noble gas configuration in its outer shell. An atom with one electron less than sodium is the highly reactive gas fluorine. Fluorine readily attracts an electron from another atom to form an **anion** to achieve a **noble gas configuration**.

He	$1s^2$
Ne	$1s^2 2s^2 2p^6$
Ar	$1s^2 2s^2 2p^6 3s^2 3p^6$
Kr	$1s^2 2s^2 2p^6 3s^2 3p^6 3d^{10} 4s^2 4p^6$
Xe	$1s^2 2s^2 2p^6 3s^2 3p^6 3d^{10} 4s^2 4p^6 4d^{10} 5s^2 5p^6$
Rn	$1s^2 2s^2 2p^6 3s^2 3p^6 3d^{10} 4s^2 4p^6 4d^{10} 4f^{14} 5s^2 5p^6 5d^{10} 6s^2 6p^6$

The noble gas configuration is where there is a completely filled energy shell and the next available higher energy level is an s shell. This is true for He where the $1s$ shell is filled and the next energy level is the $2s$ shell. For the other noble gases, we have completely filled p shells and the next highest energy level is the s sub-shell for the next quantum number n. There is a large energy gap between a p sub-shell and the next highest s shell and this gives the noble gas elements a high **ionisation energy**. Because all the occupied energy shells are filled, the electric charge distribution within noble gas elements is symmetric, resulting in no external electric field. The total angular momenta within these atoms adds up to zero, resulting in no external magnetic fields. These elements find it very difficult to form bonds with other atoms because they have little opportunity to interact electrostatically, magnetically, or energetically.

3.7 Atomic Size

The size of an atom cannot be precisely stated since the location of the electrons can only be described statistically. **Atomic size**, for practical purposes, can be defined as the most probable distance from the nucleus to the most outer-shell electron but depends on whether the atom is free or combined with another.

Electrons are attracted to the nucleus by **electrostatic forces**, but are constrained to occupy defined orbits, or energy shells, by the principles of **quantum mechanics**. The electrostatic attraction between the nucleus and a near-shell electron is much larger than between the nucleus and an outer-shell electron, not only because of the increased distance, but also because of the **screening effect** that the inner electrons have on the outer electrons.

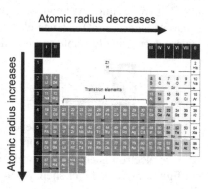

As we go across a **period** in the periodic table, the atomic number increases and so the magnitude of the positive nuclear charge also increases, and therefore so does the magnitude of the **electric field** within which the electrons find themselves. However, across any one period, the outer electrons exist in energy levels for one particular quantum number n. For example, in the 3rd period, the $3s$ and $3p$ levels are occupied. Because the electric field is also increasing as we go from left to right, so the electrons are acted upon by a larger Coulomb force and the **atomic radius** decreases across a period. However, because of the **screening effect** offered by the inner-shell electrons, the decrease in size is not as large as expected on the basis of the nuclear charge alone.

For a given vertical **group** of elements in the periodic table, the size of atoms increases as the atomic number Z increases (as we go down the column). We may therefore expect that the atomic radii would decrease but as we go down a group, electrons are filling states with a higher quantum number n and this more than offsets the effect of increasing Z.

Therefore **atomic radius** increases as we go down a group in the periodic table.

$$r = n^2 \left(\frac{\varepsilon_o h^2}{\pi m_e Z q_e^2} \right)$$

The Bohr model of the atom shows how the atomic radius increases in proportion to n^2 and decreases (inversely) with Z.

3.8 Covalent Radii

When atoms form a **covalent bond**, the nuclei of the two atoms move close together (see Section 1.7) compared to two isolated atoms adjacent to each other.

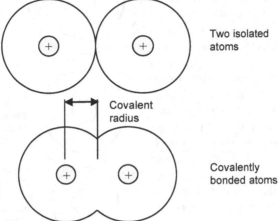

Two isolated atoms

Covalent radius

Covalently bonded atoms

The **covalent radius** of an atom is defined as one half the distance between the nuclei of two identical *covalently bonded* atoms. The actual covalent radius of an atom in a molecule where the atoms are not identical depends on the atoms involved. The general term "**atomic radius**" usually means the covalent radius (where the two atoms are identical) and this serves as a useful benchmark for comparison with other atomic sizes.

For metallic atoms, the atomic or **metallic radius** is usually defined as half the distance between two nuclei of the atoms in the solid metallic state.

The shapes of the orbitals in which the valence electrons participate in bonding is important. The s orbital has a somewhat spherical shape centred on the nucleus, while the p orbitals have a dumbbell shape along the three coordinate axes. The shapes of the atomic orbitals involved in the formation of a particular bond determine the shape of the molecule.

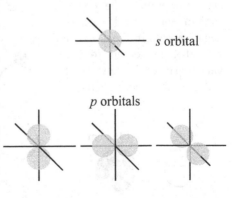

s orbital

p orbitals

3.9 Ionic Radii

When a metal atom loses an electron to become a positively charged **ion**, a reduction in **atomic size** occurs. The electron that is lost is usually an outer-shell electron and so this electron shell becomes vacant and so the outermost electron shell is the next one closer in towards the nucleus. However, the reduction in size is not just due to an outer electron shell becoming vacant. When a positive ion is formed, the remaining electrons can bunch together a little more closely because of the reduced degree of mutual repulsion (since there are now fewer electrons surrounding the nucleus).

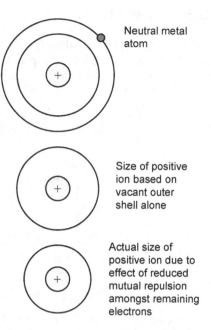

Neutral metal atom

Size of positive ion based on vacant outer shell alone

Actual size of positive ion due to effect of reduced mutual repulsion amongst remaining electrons

When a non-metal atom gains an electron to become a negative ion, this usually results in an outer-level electron shell accepting an additional electron – which would not ordinarily cause an appreciable increase in size, but the addition of this electron means that there is now an excess of negative charge in the outer shell and mutual repulsion causes these electrons to move apart and the atom "expands".

In a sodium chloride ionic crystal, therefore, the positive sodium ions shrink and the negative chlorine ions expand such that the sodium ions are almost packed in between the spaces of the chlorine ions.

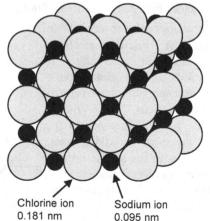

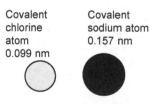

Covalent chlorine atom 0.099 nm

Covalent sodium atom 0.157 nm

Chlorine ion 0.181 nm

Sodium ion 0.095 nm

3.10 Ionisation Energy

The **ionisation energy**, or **ionisation potential**, is the amount of energy needed to remove an electron from an isolated atom. The first ionisation energy is that needed to remove one outermost electron from an atom. Successive ionisation energies are those needed to remove further electrons from the atom.

Successive ionisation energies always increase at each electron removal. As each electron is removed, the size of the atom decreases markedly and so the remaining electrons are closer to the positively charged nucleus, and so more energy is needed to remove them. Further, a particularly large increase in energy is needed to remove an electron from a filled energy level because a noble gas configuration of electrons is extremely stable.

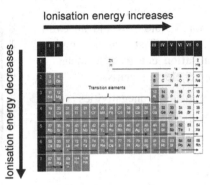

Ionisation energy increases

Ionisation energy decreases

Within a horizontal **period**, there is an increase in nuclear charge. As the **nuclear charge** increases (number of protons on the nucleus), the magnitude of the electric field within which the electrons find themselves also increases, and so generally speaking, the ionisation energy increases within a period. Also, the **atomic size** decreases from left to right and so the electrons for atoms towards the right are closer to the nucleus compared to those on the left, and so this also results in a general increase of **ionisation energy** from left to right.

Within a vertical **group**, there is generally a decrease of ionisation energy from top to bottom which is predicted on the basis of size of the atom alone. Any potential increase in ionisation energy on the basis of increased nuclear charge, going from top to bottom, is essentially cancelled out by the increased **screening effect** of the inner electrons on the outer electrons.

Generally speaking, elements with a low ionisation energy are on the left-hand side of the periodic table and those with a high ionisation energy are on the right. Noble gas configurations (eight valence electrons) can cause exceptions to these general rules.

3.11 Electronegativity

The **electronegativity** describes the relative ability for an atom, when it combines with another atom, to become "more negative" by more strongly capturing a shared electron or electron pair.

It is found that the more difficult it becomes for an atom to lose an electron, the easier it is for the atom to add an extra electron. That is, a higher **ionisation energy** also means a higher electronegativity.

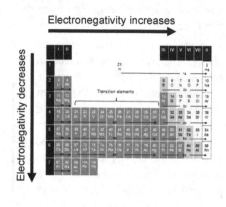

Some electronegativities within a **group** are:

Be	1.5
Mg	1.3
Ca	1.0
Sr	0.95
Ba	0.9
Ra	0.9

Some electronegativities within a **period** are:

Li	Be	B	C	N	O	F	Ne
1.0	1.5	2.0	2.5	3.0	3.5	4.0	-

The electronegativity of fluorine is so strong that in some circumstances, it is able to attract an electron from helium and so produce a compound that involves a noble gas.

The electronegativity of an atom influences the nature of the bonding between different types of atoms. Even when a bond is covalent, the shared electron spends more time nearer to the atom which has a higher electronegativity. The resulting molecule becomes polar.

4. Chemical Bonds

Summary

Lewis electron
dot structures

$$: \overset{\cdot\cdot}{\underset{\cdot\cdot}{X}} :$$

Van der Waals

Hydrogen bond

Oxidation numbers:

a) In free, uncombined, elements, the oxidation number
    of each atom is set to 0.

b) In compounds involving hydrogen, the oxidation
    number of hydrogen is 1+ .

c) In compounds involving oxygen, the oxidation
    number of O is usually 2−.

d) The sum of all the oxidation numbers of all atoms in
    an ion is equal (in both magnitude and sign) to the
    charge on the ion.

e) The sum of all the oxidation numbers of all atoms in
    a neutral molecule is 0.

Oxidation → Increase in Oxidation No. → Loss of Electrons

Reduction → Decrease in Oxidation No. → Gain of Electrons

4.1 Chemical Bond

Consider two hydrogen atoms that are initially a long way apart from each other. The two atoms are both electrically neutral, and have no ionic character. Why should they then be attracted to each other and form a stable H_2 molecule? Atoms tend to arrange themselves in the lowest possible energy state. In a **covalent bond** (like in H_2), the sharing of electrons provides the necessary reduction in total energy since the shared electrons, now having more space to move (over two atoms instead of one) have a lower **energy** than a single electron orbiting a single nucleus (see Section 1.5). This reduction in energy appears to us like a force of attraction acting between the atoms.

When two atoms form a **chemical bond**, they take up equilibrium positions according to the balance between long-range attractive and short-range repulsive forces $F(r)$ between them. At the **equilibrium position**, the potential energy $V(r)$ of the bond is a minimum.

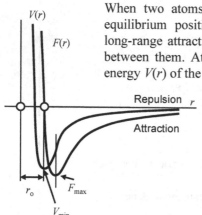

The potential energy function ($V(r)$) acting between two atoms can be very complex. Simple models are usually used, a very popular one being the **Lennard-Jones potential**:

$$V(r) = -\frac{A}{r^6} + \frac{B}{r^{12}}$$

In this potential, the negative term is the energy associated with the attractive forces and the positive term is the energy associated with the repulsive forces.

It is convenient to assign a potential energy of zero to widely spaced molecules or atoms so that when they approach, the potential energy becomes more negative. Since we generally assign a positive number to **work** done on a system (energy entering a system) and a negative number to work done by a system (energy leaving a system), work has to be done on the molecules or atoms to separate them. Thus, when the atoms are at the equilibrium position, energy is required to be done on the system to move one atom from the equilibrium distance to infinity against the force of attraction. This energy is called the **bond energy** (usually in kJ/mol).

To be consistent with the assignment of energy, we must therefore treat **attractive forces** as negative and repulsive forces as positive. The force F, with this sign convention, when multiplied by the distance of movement r, then results in an energy of the correct sign.

4.2 Lewis (Electron Dot) Formulae

Molecules and compounds form when bonds exist between atoms involving the transfer or sharing of valence electrons. The **Lewis symbol** for an atom is the element symbol surrounded by dots that represent the possible **valence electrons** for the atom.

$$: \overset{\displaystyle ..}{\underset{\displaystyle ..}{X}} :$$

Lewis recognised that bonding or sharing of electrons occurred so that an element tended to acquire eight valence electrons (noble gas configuration): the **octet rule.** To represent a bond between two atoms, a **Lewis structure** is written so as to indicate the position of the valence electrons:

$$H \cdot + \cdot \overset{\displaystyle ..}{\underset{\displaystyle ..}{Cl}} : \longrightarrow \left(H \overset{\displaystyle ..}{\underset{\displaystyle ..}{Cl}} : \right)$$

This **Lewis structure** shows how each element acquires a noble gas configuration when a chemical (in this case, ionic) bond is formed.

Group	Lewis formula
I	E
II	E ·
III	E ·
IV	· E ·
V	· E ·
VI	· E :
VII	· E :

"E" denotes an element.

When determining a Lewis structure, the total number of dots that appear (the total number of **valence electrons**) is the sum of the number of valence electrons for each participating atom.

In some cases, it is not easily determined what the Lewis or **electron dot** formula for an atom should be. For example, in carbon, the electronic configuration is: $1s^2\ 2s^2\ 2p^2$ and we might be tempted to write the **electron dot** formula as:

$$: \overset{\displaystyle \cdot}{C}$$

Although this arrangement can occur, when carbon usually combines with other atoms, the $2s$ orbitals combine with the $2p$ orbitals and the electrons are spread out over the resulting combined "sp" energy level giving four valence electrons. Carbon is **tetravalent.**

$$\cdot \overset{\displaystyle \cdot}{\underset{\displaystyle \cdot}{C}} \cdot$$

4.3 Multiple Bonds

Covalent bonds between atoms often involve sharing of more than one electron at a time. For example, carbon has a bonding capacity, or valence, of four, and so the carbon compound **ethane** is a stable molecule and is fairly unreactive.

$$
\begin{array}{cc}
\text{H} & \text{H} \\
\text{H} : \ddot{\text{C}} : \ddot{\text{C}} : \text{H} \\
\text{H} & \text{H}
\end{array}
$$

Ethylene, with molecular formula C_2H_4 should therefore leave two lone valence electrons:

$$
\begin{array}{cc}
\text{H} & \text{H} \\
\text{H} : \ddot{\text{C}} : \ddot{\text{C}} : \text{H} \\
\cdot & \cdot
\end{array}
$$

We would therefore expect ethylene to be a very reactive compound, in much the same way as Group I elements. However, this is not the case. Instead, the two carbon atoms arrange for the orbits of the two lone electrons to overlap and be shared. That is, two pairs of electrons are shared between the carbon atoms to form a **double bond**.

$$
\begin{array}{c}
\text{H} : \text{C} :: \text{C} : \text{H} \\
\ddot{\text{H}} \quad \ddot{\text{H}}
\end{array}
$$

   Triple bonds are also possible whereby three pairs of electrons are shared. The double and triple bonds offer even a greater range of movement of valence electrons (since more electrons are shared and so more electrons have more overall room to move) compared to a single bond. The strength of the bond is higher, and the C nuclei are closer together (shorter bond length).

   Although the bond energy associated with a double bond is higher, the double bond results in a more reactive compound than a comparable single-bonded structure. This is because the double bond offers more opportunities for other more reactive elements than carbon to attach to these valence electrons – as if other elements can open up the bond and find two potential bonding sites rather than one. It is energetically more favourable to have two single bonds than one double bond.

Molecule	Structural formula	C – C length (Å)	C – C energy (kJ/mol)
C_2H_6 Ethane	$\begin{array}{c}\text{H} \quad\quad \text{H}\\ \text{H}-\!\!>\!\!\text{C}\!-\!\text{C}\!<\!\!-\text{H}\\ \text{H} \quad\quad \text{H}\end{array}$	1.54	334
C_2H_4 Ethylene	$\begin{array}{c}\text{H} \quad\quad \text{H}\\ \text{C}\!=\!\text{C}\\ \text{H} \quad\quad \text{H}\end{array}$	1.33	606
C_2H_2 Acetylene	$\text{H}-\text{C}\equiv\text{C}-\text{H}$	1.20	828

4.4 Lewis Single-Bonded Structures

Some examples of single-bonded structures are:

Molecular formula	Total No. valence electrons	Lewis formula	N	Shape	Note
F_2 Fluorine gas	14	:F̈:F̈:	-	F — F	Linear
CH_4 Methane	8	H:C̈:H (with H above and H below)	4	H–C(H,H)–H	Tetrahedral
$SiCl_4$	32	:C̈l: / :C̈l:S̈i:C̈l: / :C̈l:	4	Cl–Si(Cl,Cl)–Cl	Tetrahedral
NH_3 Ammonia	8	H:N̈:H (with H below)	4	H–N(H,H)	Pyramidal
H_2O Water	8	:Ö:H (with H below)	4	H–O–H	Bent
NH_4^+	8	H:N:H (with H above and H below)	4	H–N(H,H)–H	Tetrahedral
C_2H_6 Ethane	14	H:C̈:C̈:H (with H H above and H H below)	4,4	H,H–C—C–H,H	2 tetrahedral
H_2O_2	14	H:Ö:Ö:H	4,4	H–O—O–H	

Note that **lone pairs**, being closer to the central atom, have a greater repulsive effect than bonded pairs and so influence the shape of the molecule. Note that each atom is surrounded by eight electrons (except for H which requires two electrons for noble gas configuration).

4.5 Lewis Multiple-Bonded Structures

Examples of multiple-bonded structures:

Molecular formula	No. valence electrons	Lewis formula	N	Shape	Note
N_2 Nitrogen gas	10	N⋮⋮⋮N	-	N ≡ N	Linear
C_2H_4 Ethylene	12	H:C::C:H H H	3,3	$\ce{C=C}$ with H's	2 Tetrahedral
C_2H_2 Acetylene	10	H:C⋮⋮⋮C:H	2,2	H—C≡C—H	Linear
CO_2 Carbon dioxide	16	:Ö::C::Ö:	2	O = C = O	Linear
HCN Hydrogen cyanide	10	H:C⋮⋮⋮N	2	H—C≡N	Linear
$COCl_2$ Phosgene	24	:Cl: C⋮⋮⋮O: :Cl:	3	Cl Cl C=O	Planar
H_2CO Formaldehyde	12	H C::Ö: H	3	H H C=O	Planar

Note that each atom is surrounded by eight electrons (except for H which requires two electrons for noble gas configuration).

4.6 Lewis Exceptions to the Octet Rule

Examples of structures that do not obey the **octet rule** include:

(1) Atoms having more than eight electrons involved in bonding:

Molecular formula	No. valence electrons	Lewis formula	N	Shape	Note
PCl_5	40		5	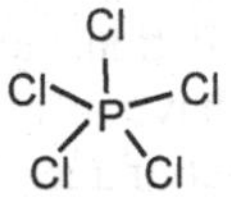	More than an octet
SF_4	43		5		More than an octet with lone pair

When there are more than eight electrons involved, such as in PCl_5, the bonding usually involves electrons from the d sub-shell.

(2) Atoms having less than eight electrons involved in bonding:

BeH_2	4	H : Be : H	2	H — Be — H	Less than an octet
NO	11	:N :: Ö:		N = O	Less than an octet

When more than one valid structure can be written and the true structure cannot be written, the actual or true structure is called a **resonance hybrid** of the different structures. An example is NO. The unpaired electron (as shown above) can be written as attached to either the N or the O atom.

Example of **resonance structure**: CO_3^{2-}

4.7 Oxidation Number

The **oxidation number**, or **oxidation state**, is a quantity that describes the number of electrons that appear to be gained or lost by an atom when a chemical bond is formed. That is, it is essentially the number of electrons that participate in the formation of the chemical bond. If we imagine all bonds to involve a transfer of electrons (even if in reality the bond may be covalent), the oxidation number is essentially the electronic charge that an atom appears to have after the bond has formed. The oxidation number is determined by a set of agreed rules. It is not strictly a physical quantity.

In a molecule or a compound, electrons that are shared between atoms are counted as belonging to the more **electronegative** atom. Electrons that are shared between two identical atoms are equally divided between the two atoms.

(a) In free, uncombined elements the oxidation number of each atom is set to zero. For example, the oxidation number of H in H_2 is zero.

(b) In compounds involving hydrogen, the oxidation number of hydrogen is 1+ except in the case of metal hydrides (where the hydrogen is bonded to an atom which is less electronegative) where it is 1–.

(c) In compounds involving oxygen, the oxidation number of O is usually 2–. The exceptions are the oxygen atom in peroxides and when oxygen is bonded with fluorine.

(d) The sum of all the oxidation numbers of all atoms in an ion is equal (in both magnitude and sign) to the charge on the ion.

(e) The sum of all the oxidation numbers of all atoms in a neutral molecule is 0.

Consider the compound H_2O. The oxygen atom, being more **electronegative** than the hydrogen, is seen to gain two electrons and so is given an oxidation number of 2– (since it seems to have acquired a net "negative" charge). Each hydrogen atom is seen to lose an electron and so the oxidation number of hydrogen in this compound is assigned as 1+.

When the oxidation number of an atom has increased, the atom is said to have been **oxidised** (as in the hydrogen above). This usually involves the loss of electrons. When the oxidation number of an atom has decreased, the atom is said to have been **reduced**, that is, its oxidation number has decreased. This usually involves a gain of electrons (as in the oxygen above).

The oxidation number is useful when naming some compounds. Prefixes like *hypo* and *per*, and suffixes like *ous, ic, ite*, are assigned according to the oxidation number of the central metal or non-metal atom in a compound.

4.8 Oxidation Number Examples

Example	Element	Oxidation state	Rule
S	S	0	a
H_2S	S	-2	b, e
Fe^{2+}	Fe	+2	d
Fe^{3+}	Fe	+3	d
MnO_4^-	Mn	+7	c, e
Mn^{2+}	Mn	+2	d
$Cr_2O_7^{2-}$	Cr	+6	c, e
Cr^{3+}	Cr	+3	d
NO_3^-	N	+5	c, e
NO	N	+2	c, e
NO_2	N	+4	c, e
SO_4^{2-}	S	+6	c, e
CO_2	C	+4	c, e
H_2O_2	O	-1	special case

When a substance is **oxidised**, it can be done so by either adding oxygen or removing hydrogen. Similarly, when a substance is **reduced**, it can be done so by either removing oxygen or adding hydrogen. For example, consider the production of methane CH_4 from carbon and hydrogen:

$$C + 2H_2 \rightarrow CH_4 \qquad \text{Oxidation No.}$$
$$2H_2 \rightarrow 4H^+ + 4e^- \qquad \text{0 to 4+}$$
$$C + 4e^- \rightarrow C^{4-} \qquad \text{0 to 4-}$$

From the point of view of the hydrogen, it has lost electrons and so is oxidised. From the point of view of the carbon, it has gained electrons and so is reduced. Of course in this compound, CH_4, the electrons are shared between the C and H atoms as covalent bonds, but from the perspective of oxidation numbers, the electrons are effectively transferred from H to C.

In some biochemical reactions, electrons (in the company of protons H^+) are transferred during oxidation and reduction reactions as a means of passing energy from one molecule to another. The addition of an electron via the addition of hydrogen stores energy in the compound being reduced. The energy is released when the compound is oxidised.

When CH_4 is combined with oxygen (e.g. when it is burnt), the C is oxidised to CO_2 and water:
$$CH_4 + O_2 \rightarrow CO_2 + 2H_2O$$

4.9 Polar Bonds

Although bonds between atoms may be classified as being **covalent** or **ionic**, in many different substances, there is a gradation from one type to the other. Unequal sharing of electrons in a covalent bond gives the bond an ionic character. Further, such a covalent molecule may acquire regions of net positive and negative charge and become **polarised**.

Separation of charges within a covalent molecule usually results in electrons spending more time on the atom which has the greatest **electronegativity**.

$$: C \; ::: \; O \; :$$

In CO, the shared electrons spend more time near O than C and so the centre of positive charge is offset from the centre of negative charge.

For example, in the molecule carbon monoxide, there is a difference in electronegativity between carbon (2.5) and oxygen (3.5) and so the shared electrons spend more time near the oxygen atom than the carbon atom. This results in a net separation of centres of **electric charge** within the molecule. The molecule is thus polarised and is a **dipole**. The molecule as a whole remains electrically neutral, but the distribution of charge within the molecule shifts so that the geometrical centre of positive charge is different than the centre of negative charge.

Although it is straightforward to determine if a diatomic molecule is polar or non-polar, by a consideration of the electronegativities of their atoms, the situation is more complicated for molecules with more than two atoms. For example, in CO_2, the individual bonds between the carbon and the oxygen molecules may be polar, but the molecule as a whole is non-polar because of the symmetry of the shape of the molecule.

$$: \ddot{O} :: C :: \ddot{O} : \qquad O = C = O \qquad \mu = 0 \; \text{debye}$$

When CO_2 is placed in an electric field there is no net **dipole moment**.

Experiments show that **water** has a permanent dipole moment due to the presence of the electron pairs on the oxygen.

$\mu = 1.84$ debye

Bond angle 104.5°

Polarization is responsible for **physical bonds** forming between molecules which in turn result in molecules forming liquids and solids.

4.10 Hybrid Orbitals

In a covalent bond, when valence electrons are shared between atoms, they do so by overlapping orbitals. Consider an oxygen atom: $1s^2$, $2s^2$, $2p^4$

$$1s \quad 2s \qquad 2p$$

oxygen $\boxed{\uparrow\downarrow}\;\boxed{\uparrow\downarrow}\;\boxed{\uparrow\downarrow\;\downarrow\;\downarrow}$

The shapes of molecules are difficult to predict on the basis of the orbitals of isolated atoms because the nature of these orbitals change when there are more than one electron in an atom and when atoms come together to form molecules.

When a molecule of water forms, the $2s$ and the three $2p$ orbitals in the oxygen combine to form a new molecular orbital called the sp^3 **hybrid**. The superscript 3 indicates that three of the original p orbitals are involved in the hybrid orbital. It is this hybrid orbital $2sp^3$ that combines with the s orbitals of the H atoms.

Hybrid	Shape
sp	linear
sp^2	plane triangular
sp^3	tetrahedral

$$1s \qquad 2sp^3$$

oxygen
sp^3 hybrid $\boxed{\uparrow\downarrow}\;\boxed{\uparrow\downarrow\;\uparrow\downarrow\;\downarrow\;\downarrow}$

Each of the $2sp^3$ hybrid orbitals is at the same energy level.

When hybridization occurs, the rearrangement of electrons often results in there being more unpaired valence electrons than there were originally. For example, in carbon compounds, we have:

$$1s \quad 2s \qquad 2p$$

carbon
(single atom) $\boxed{\uparrow\downarrow}\;\boxed{\uparrow\downarrow}\;\boxed{\downarrow\;\downarrow\;\;}$

$$1s \quad 2sp^2 \quad 2p$$

sp^2 hybrid $\boxed{\uparrow\downarrow}\;\boxed{\downarrow\;\downarrow\;\downarrow}\;\boxed{\downarrow}$

In CH_4, it is the hybridization of the C atomic energy levels into four identical $2sp^3$ orbitals that results in a regular tetrahedral geometry with bond angle 109.5°. In PCl_5 (more than eight electrons bonded), the hybrid orbitals are sp^3d.

$$1s \qquad 2sp^3$$

sp^3 hybrid $\boxed{\uparrow\downarrow}\;\boxed{\downarrow\;\downarrow\;\downarrow\;\downarrow}$

In the double C=C bond of C_2H_4 one of the bonds is the end-to-end overlap of one of the sp^2 hybrid orbitals (this is called the σ **sigma bond**) along the line of the two C atoms (the other two sp^2 orbitals overlap with the s orbitals of the H atoms), while the other C bond is formed by the side-by-side overlap of the unhybridised (dumbbell shaped) $2p$ orbitals parallel to the plane of the C and H atoms (this is called the π **pi bond**).

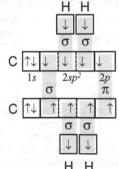

4.11 Polarisation

When an electric field E is applied to a molecule in a non-polar insulating material, the centre of charge of electrons moves left, the centre of charge of protons move right and the two centres of charge are then separated by distance d. The atom or molecule becomes **polarised** in the presence of E.

Polarisation in a material may occur due to several mechanisms, all of which may occur to some extent depending on the atom. Most often, polarisation occurs when an external electric field E is applied.

1. **Electronic polarisation**

   Small induced dipole moment arising from difference in the net centres of nucleus and electrons in an atom.

   $\varepsilon_r \approx 2 - 4$

   The entities with circles around them are **dipoles**.

   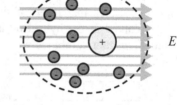
   E

2. **Ionic polarisation**

   Dipole moment created by shift of positive ions with respect to negative ions in unit cell.

   $\varepsilon_r \approx 6 - 10$

   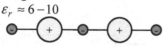
   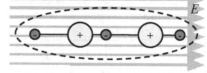

   When field is applied, movement of ions produces a net dipole moment in a unit cell and hence a net polarisation.

3. **Dipolar polarisation**

   Net dipole moment created by alignment of molecule with external field due to presence of internal permanent dipoles from geometrical structure of molecule.

   $\varepsilon_r \approx 20 - 100$

   Water molecule has a permanent dipole moment.

   CO_2 molecule has no permanent dipole moment. No dipolar polarisation can occur.

   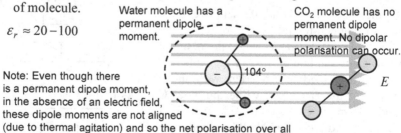

   Note: Even though there is a permanent dipole moment, in the absence of an electric field, these dipole moments are not aligned (due to thermal agitation) and so the net polarisation over all molecules in the material is zero. When a field E is applied, molecular dipoles tend to align themselves with the field and there exists a net polarisation.

4.12 van der Waals Forces

The arrangement of electrons around atoms is not constant in time. Electrons move within their energy shells or levels with kinetic energy. An instantaneous movement of an electron within an atom may result in the atom becoming **polarised** for a short period of time. During this time, the polarisation in one atom

may induce polarisation in a neighbour atom, resulting in a very brief electrostatic attraction between them. These instantaneous forces of

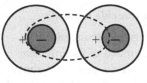

Instantaneous polarisation of one molecule results in induced dipole in neighbour molecule and a net instantaneous attraction between the two molecules.

attraction and repulsion are termed **London forces** after the scientist who proposed the explanation for them and are found in all covalent materials.

In materials containing molecules that have permanent dipole moments, the molecules tend to align themselves (against the tendency of thermal agitation to keep them in random orientation). The electrostatic attraction between the polarised ends of the molecules draws them together. **Dipole-dipole interaction** may persist over many molecules and appears to us like a force holding the molecules together.

When considered over millions of atoms, the London and permanent dipole forces can be substantial and are called physical, as distinct from chemical, bonds between molecules. The intermolecular forces of this type are called **van der Waals forces**. In most cases, it is the van der Waals forces that bind atoms and molecules into liquids.

Intermolecular forces on this scale are responsible for deviations from ideal behaviour in liquids and gases. Van der Waals forces operate over a small scale in comparison to distances between molecules in a gas. It is only when gas molecules become closer together (such as under increased pressure) that these forces become strong enough to cause the molecules to then clump together and **condense** into a **liquid**. The forces become less effective when the temperature increases due to the tendency for an increase in random order associated with thermal agitation of molecules.

4.13 Hydrogen Bond

Hydrogen has one valence electron and so can readily form a single covalent bond. However, there are some compounds involving hydrogen in which the hydrogen atom can be said to be bonded to two atoms at the same time. Such compounds involve hydrogen bonding with small electronegative atoms.

Consider the **hydrogen fluoride** molecule:

Because fluorine has a large **electronegativity**, the shared electron spends much of its time near the fluorine atom. The molecule, as shown above, is **polar**. Now, if there are other HF molecules nearby, then the positively charged H side of the molecule will be drawn into contact with the negatively charged F end of the other molecule. Two molecules thus join up so:

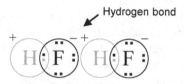

It looks to us like the H atom in the middle is holding the two outer F atoms in place like a tie bar. This electrostatic attraction is called a **hydrogen bond**. The hydrogen bond can only form between hydrogen and small electronegative atoms because the H "ion" is very small and only has space for one other atom to be located near enough to its polar end to form an **electrostatic** bond.

Hydrogen bonds also form between **water molecules**. Hydrogen bonding between water molecules gives rise to a characteristic crystal structure in **ice**, which in turn is reflected in the shape of **snowflakes**.

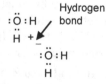

Hydrogen bonding is very important in biological processes. The bonds, having an energy of about 20 kJ, are stronger than those due to van der Waals forces (about 0.2 kJ) but are significantly weaker than those in chemical bonds (about 200 kJ). Certain biological processes rely on these bonds being of just the right strength for events to occur with ease (such as in the forming and breaking of the two halves of the molecular spiral in **DNA** – these hydrogen bonds involve H, O and N atoms).

5. States of Matter

Summary

Gases: $\dfrac{p_1 V_1}{T_1} = \dfrac{p_2 V_2}{T_2}$ Combined gas law

$p_1 V_1 = p_2 V_2$ Boyle's law

$\dfrac{V_1}{T_1} = \dfrac{V_2}{T_2}$ Charles' law

Volume ⌐ ┌Absolute
                    temperature

$$pV = nRT$$

Absolute ── ↑ ↑└ Universal gas
pressure   no. moles constant
                       8.3145 J/mol/K

Liquids: $\text{Molarity} = \dfrac{\text{No. moles of solute}}{\text{Volume of solution in litres}}$

Solids:

- Molecular solids - in which the binding forces between discrete atoms or molecules are due to van der Waals interactions between instantaneous electric dipoles.

- Ionic solids - in which the binding energy is due to Coulomb electrostatic forces between positively and negatively charged ions in the crystal.

- Covalent solids - in which the binding energy is due to shared valence electrons between atoms in the solid.

- Metallic solids - in which valence electrons are effectively shared amongst all the atoms in the solid.

- Amorphous solids – no long-range structure.

5.1 Changes of State

When a solid is heated, atoms/molecules within the solid absorb heat energy and this is stored as **internal energy** of vibration. As the temperature increases, the magnitude or amplitude of the vibration increases until the bonds holding the molecules together become temporarily ruptured and molecules flow over one another. This is called **melting**.

At the **melting point**, further heating results in no increase in temperature. The energy input in this case goes to breakage of bonds between the molecules and the material changes phase from solid to liquid.

Upon further heating, molecules of liquid increase their translational energy until they acquire sufficient kinetic energy to escape the liquid into the gaseous phase.

The presence of impurities lowers the melting point. A measurement of the melting point can be used as an index of the purity of a particular substance.

When a liquid cools, molecules begin to lose their internal energy of motion until eventually, intermolecular forces are able to hold them in place as a solid. The temperature at which this occurs is called the **freezing point** – which is usually equal to the melting point.

If the molecules in the cooling solid do not immediately lock into the pattern of the solid, then the temperature may fall below the official freezing point and the material remains in the liquid phase. This is called **supercooling**. Introduction of a seed crystal may cause the liquid to avoid the supercooling condition.

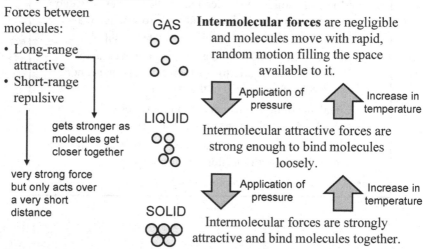

Forces between molecules:

- Long-range attractive
- Short-range repulsive

gets stronger as molecules get closer together

very strong force but only acts over a very short distance

GAS **Intermolecular forces** are negligible and molecules move with rapid, random motion filling the space available to it.

Application of pressure

Increase in temperature

LIQUID Intermolecular attractive forces are strong enough to bind molecules loosely.

Application of pressure

Increase in temperature

SOLID Intermolecular forces are strongly attractive and bind molecules together.

5.2 Changes of State of Liquids/Gases

Closed system

Open system

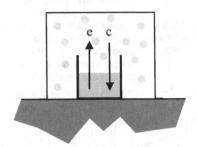

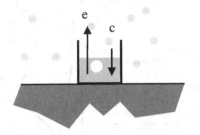

A dynamic equilibrium is set up where the rate of evaporation = the rate of condensation.

The **equilibrium vapour pressure (EVP)** or **saturated vapour pressure** is the pressure exerted by the vapour when in equilibrium with its liquid. The magnitude of the EVP depends on the nature of the liquid and the temperature.

- Nature of liquid. The stronger the **van der Waals forces** between molecules, the lower is their ability to escape the liquid and the lower the EVP.

- Temperature. Increase in temperature increases kinetic energy of molecules in the liquid and hence their ability to escape from the surface and hence the higher the EVP.

A liquid is said to boil at a temperature called the **boiling point**, at which the EVP of the liquid becomes equal to the prevailing atmospheric pressure. Vapour bubbles form in the liquid and rise to the surface.

The **normal boiling point** is the boiling point at 1 atm of pressure (101.3 kPa).

Increasing the prevailing atmospheric pressure increases the boiling point.

5.3 Phases of Matter

In a **p–V diagram**, the temperature is kept constant, volume decreased and pressure recorded.

Critical Temperature T_c
There is, for each gas, a temperature above which the attractive forces between molecules are not strong enough to produce liquefaction no matter how high a pressure is applied.

T_c H_2O = 647K at 218 atm
T_c He = 5.2K at 2.3 atm

Vapour pressure
The partial pressure exerted by the vapour when it is in equilibrium with its liquid. It depends on the temperature and nature of the substance. The temperature at which the vapour pressure equals the prevailing atmospheric pressure is called the **boiling point**.

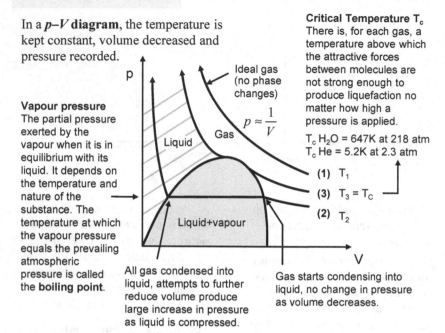

(1) T_1
(3) $T_3 = T_c$
(2) T_2

$p \approx \dfrac{1}{V}$

All gas condensed into liquid, attempts to further reduce volume produce large increase in pressure as liquid is compressed.

Gas starts condensing into liquid, no change in pressure as volume decreases.

In a **phase diagram**, we keep the volume V constant and plot pressure versus temperature.

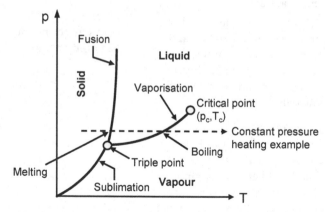

At each point (p,T) only a single phase can exist except on the lines where there is phase equilibrium. At the **triple point**, solid, liquid and vapour exist together in **equilibrium**.

5.4 Gases

The properties of a perfect or ideal gas are the most convenient to study. The properties of an ideal gas are:

- The molecules of the gas occupy a very small volume compared to the volume of the container.
- The molecules are very distant from one another and only interact elastically during collisions.

Real gases often behave like ideal gases at conditions in which no phase changes occur.

Macroscopic properties of a gas:

- Pressure
- Temperature
- Volume
- Mass

⎫ These quantities specify the **state** of a gas

Consider a mass (m) of gas:

If p, V and T all vary, then:

$$\frac{p_1 V_1}{T_1} = \frac{p_2 V_2}{T_2}$$ **Combined gas law**

If temperature T is a constant:

$$p_1 V_1 = p_2 V_2$$ **Boyle's law**

If pressure p is a constant:

$$\frac{V_1}{T_1} = \frac{V_2}{T_2}$$ **Charles' law**

Note: These laws cannot be applied when the mass of gas changes during the process. Pressures and temperatures are absolute.

No. moles ⟶ $n = \dfrac{m}{M_m}$ ⟵ Mass in kg, Molar mass

Let us express the mass of a gas indirectly by specifying the number of moles. Experiment shows that Boyle's law and Charles' law lead to:

By using moles, we get the ideal gas equation with the universal gas constant R (units J/mol/K). Otherwise, value of R depends on the nature of the gas (i.e. no longer universal) and has units J/kg/K.

Volume ⟶ ┌Absolute temperature

$$pV = nRT$$

Absolute pressure ┘ └ Universal gas constant

No. moles   8.3145 J/mol/K

Example: Calculate the volume occupied by one mole of an ideal gas at 273 K at atmospheric pressure.

$$pV = nRT$$
$$101.3(V) = 1(8.314)(273)$$
$$V = 22.406\,\text{L}$$

This equation links all the macroscopic quantities needed to describe the (steady) state of an *ideal* gas and is thus called an **equation of state**.

5.5 Solutions

Homogenous mixtures of two or more gases, liquids, or solids are called
solutions.

- Gases usually mix in any proportion. Air is a solution of predominantly
  nitrogen, oxygen and carbon dioxide.

- Liquid solutions can be made by dissolving a solid, liquid or gas (the
  solute) in a liquid (the **solvent**). When the solvent is water, the solution is
  called an **aqueous solution**.

- Solid solutions occur where atoms of one solid are randomly dispersed
  throughout the other solid. **Alloys** are common examples of solid
  solutions although some alloys are **compounds** (combine in definite
  proportions). Some alloys are also heterogeneous **mixtures**.

Intermediate between heterogeneous mixtures and homogenous solutions
is the case where the dispersed solid particles are not broken up into
molecular units and yet are not in collections large enough to be called a
separate phase. In this case, the dispersion is called a **colloid**.

The strength, or **concentration**, of a solution can be described as:

- The ratio of the number of moles of one of the components (the solute or
  the solvent) relative to the total number of moles present is called the
  mole fraction of the component.

- The number of moles of solute per litre of final *solution* is called the
  molarity and given the symbol M (this is the most common method but
  depends on volume, which in turn is temperature dependent).

- The number of moles per kilogram of *solvent* is called the **molality** and is
  given the symbol m (has the advantage of not being dependent on
  temperature).

- The percent solute by mass or volume of the final solution.

- % by weight (**w/w**) is the number of grams of solute per 100 g of
  solution.

- % by volume (**w/v**) is the number of grams of solute per 100 mL of
  solution.

The limiting concentration of the solution to which a solute can be
dissolved in a solvent is called the **solubility**. At this concentration, the
solution is said to be **saturated**. Addition of excess solute results in no
increase in concentration and a portion of the solute remains undissolved.

5.6 Aqueous Solutions

Many chemical reactions involve chemicals that exist as water, or **aqueous solutions**.

$$\text{Molarity} = \frac{\text{No. moles of solute}}{\text{Volume of solution in litres}}$$

Examples:

(1) Calculate the mass of solute needed to make up 1 litre solution of 0.5M sodium hydroxide:

$$1 \text{ mole NaOH} = 23 + 16 + 1$$
$$= 40 \text{ g}$$

mass required for 1 L solution at $0.5M = 40(0.5)$
$$= 20 \text{ g}$$

(2) Calculate the volume of 10M acid required to make up 250 mL of 0.4M solution of HCl:

$$1 \text{ L} = 0.4\text{M} \qquad\qquad \text{Cl} = 35.5$$
$$= (0.4)(36.5)\text{g} \qquad\qquad \text{H} = 1$$
$$= 14.6 \text{ g}$$
$$0.25 \text{ L} = 3.65 \text{ g HCl required}$$
$$10\text{M} = 365 \text{ g/L HCl}$$
$$3.65 \text{ g HCl} = 10 \text{ mL}$$

It is often observed that for dilute aqueous solutions, the addition of the solute lowers the **freezing point** and raises the **boiling point** (reduction in vapour pressure) compared to pure water.

When an aqueous solution evaporates, a **hydrated ionic compound** may be formed. The hydrated compound contains water of crystallization:

$$Co^{2+}(aq) + 2Cl^-(aq) \xrightarrow{\text{evap}} CoCl_2 + 2H_2O(s)$$

5.7 Solubility of Solids

When a solute dissolves in a solvent, three distinct processes can be identified:

- Solute-solute interactions – whereby the solute atoms or molecules are separated from each other (such as the dissociation of Na^+ and Cl^- into ions). This is an **endothermic** process (heat is required to break the ionic bonds).
- Solvent-solvent interactions – whereby the solvent atoms or molecules may be separated from each other (such as when H_2O molecules are separated from one another to accommodate the incoming solute molecules). This is an **endothermic** process.
- Solute-solvent interactions – whereby processes occur between the solute and the solvent. These are usually **exothermic** (reduction in total energy).

Some ionic compounds are soluble in water, but insoluble in non-polar solvents (like benzene and carbon tetrachloride). When water is the solvent, a **hydration** process can occur which serves to insulate the positive and negative ions from each other and prevent them reforming into a solid. This is a solute-solvent interaction.

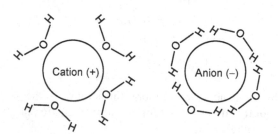

Hydration involves the formation of bonds and so is exothermic (heat is released as the total energy is lowered) and so the solute-solvent interactions dominate the solute-solute interactions. Non-polar solvents cannot hydrate the ions.

Some non-polar molecular solids are soluble in non-polar solvents but are insoluble in water (a polar solvent). Non-polar solute molecules are not hydrated by water, and so solute-solvent interactions are not significant here. If more energy is required to break the hydrogen bonds linking water molecules together than that required to break the bonds holding the components of the solute together, then the solute components remain insoluble. When placed in a non-polar solvent, solvent-solvent interactions may occur and the solvent molecules separate easily to accommodate solute molecules and the solute dissolves.

5.8 Solubility Equilibrium

If a solute is added to the solvent in increasing quantity, a point is reached where the atomic or molecular species being dissolved can no longer be accommodated in the solute. As more solute is added, the reverse process begins to occur so that some of the previously dissolved solute returns to the solute phase. This is called **precipitation**. This is most vividly illustrated when a solid solute is added to a liquid solvent. When the rate of precipitation becomes equal to the rate of **dissolution**, dynamic equilibrium is achieved and the solution is said to be **saturated**.

The **concentration** of a solute in a given solvent at **saturation** is called the **solubility** of the solute in the solvent.

Solubility depends on **temperature**. When a solute is dissolved in a solvent and heat is generated, the solubility generally decreases with increasing temperature. If heat is absorbed, the solubility generally increases with increasing temperature. Often, whether or not heat is absorbed or generated depends on the physical processes involved in any chemical reactions that may occur in the solution.

Solubility depends on **pressure**. For a non-reactive gas being dissolved in a liquid, the solubility C increases as the pressure P increases according to **Henry's law**: $C \propto P$

For example, dissolved CO_2 in a soft drink readily precipitates (observed as bubbles) when the pressure above the liquid is released (such as when the bottle is opened). An easy way to remember Henry's law is to think of it as the soft drink law.

For example, air dissolves in water to some extent and when it does so, heat is produced (since the gas "condenses" into a liquid form). The solubility of air in water thus decreases as the temperature increases. When water is heated, the dissolved air can be seen to "precipitate" as bubbles before the boiling point of water is reached.

One of the most common solvents is water, but not all compounds are soluble in water.

Compound	General rule	Exceptions
Nitrates	Soluble	
Chlorides	Soluble	$AgCl$, $PbCl_2$, Hg_2Cl_2
Sulphates	Soluble	$PbSO_4$, $BaSO_4$
Carbonates	Insoluble	Na_2CO_3, K_2CO_3, $(NH_4)2CO_3$
Hydroxides	Insoluble	$NaOH$, KOH, $Ba(OH)_2$, (NH_4OH)
Oxides	Insoluble	Na_2O, K_2O, BaO

5.9 Electrolytes

Some molecular and ionic compounds dissolve in water to give a solution that is able to conduct electric current. The compounds, when dissolved in water, dissociate into **ions**. The ions, being electrically charged, are thus able to move under the influence of an electric field and the movement of these ions is therefore an electric current.

- **Strong electrolytes** dissociate completely into ions. Many ionic compounds are strong electrolytes. Molecular compounds that are strong electrolytes are usually acids.

- **Weak electrolytes** only partially dissociate into ions. Usually, molecular compounds that are weak electrolytes are either an acid or a base.

- **Non-electrolytes** do not dissociate into ions when placed in solution. Molecular compounds that are neither acids or bases are usually non-electrolytes.

Molecular compounds that are **strong electrolytes**:

	Molecular formula	Ions formed in solution	
Hydrochloric acid	HCl	H^+, Cl^-	Strong acid
Nitric acid	HNO_3	H^+, NO_3^-	Strong acid
Sulphuric acid	H_2SO_4	H^+, SO_4^{2-}	Strong acid
Perchloric acid	$HClO_4$	H^+, ClO_4^-	Strong acid

Molecular compounds that are **weak electrolytes**:

	Molecular formula	Ions formed in solution	
Acetic acid	CH_3COOH	H^+, $CH_3CO_2^-$	Weak acid
Benzoic acid	C_6H_5COOH	H^+, $C_6H_5CO_2^-$	Weak acid
Ammonia	NH_3	NH_4^+, OH^-	Weak base

Molecular compounds that are **non-electrolytes**:

	Molecular formula	Ions formed in solution
Ethyl alcohol	C_2H_5OH	
Sucrose	$C_{12}H_{22}O_{11}$	

5.10 Osmosis

Osmosis is a movement of molecules through a **semi-permeable membrane**. A **semi-permeable membrane** is a sheet of substance that can allow some molecules through it but not others. Semi-permeable membranes can be things like **cell walls** in living tissue that contain openings that allow small molecules to pass through but not large molecules.

The concept of **osmosis** depends on the phenomenon of **diffusion**. Diffusion occurs when there is a **concentration gradient**. The concentration gradient drives the diffusion process. It is a fundamental consequence of the **second law of thermodynamics**. Diffusion happens in a wide variety of physical processes (e.g. the formation of the barrier potential at the p-n junction of a **semiconductor**).

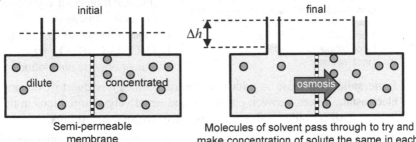

Semi-permeable membrane

Molecules of solvent pass through to try and make concentration of solute the same in each.

Consider a **dilute solution** on one side of a semi-permeable membrane and a **concentrated solution** on the other side. In this example, the semi-permeable membrane allows molecules of **solvent** to pass through it, but not the **solute**. Be careful. In this case, the **concentration gradient** has to be looked at from the point of view of the solvent. In this case, molecules of solvent tend to pass from the dilute solution into the concentrated solution so as to attempt to achieve uniformity of concentration throughout. Molecules of solute from the concentrated solution would very much like to pass into the dilute side to achieve the same aim, but cannot pass through the membrane. The movement of solvent from one side to the other results in an increase in pressure on the concentrated side. The increase in pressure tends to inhibit the further flow of solvent and an equilibrium condition is reached where no more net movement occurs across the membrane. This pressure increase is called **osmotic pressure**.

For dilute solutions, the osmotic pressure π in atm (atmospheres) of non-electrolytes is given by $\pi = MRT$ where M is the molar concentration of the non-electrolyte, T is in K, and $R = 0.0821$ L atm/mol/K.

5.11 Solids

Generally speaking, there are two kinds of solids. **Crystalline solids** possess a repeating order of atoms, ions, complex ions, or molecules. The repeating pattern is called the **crystal lattice**. The repeating unit in the crystal lattice is called a **unit cell**. **Amorphous solids** have no repeating structure of atoms, ions or molecules.

Crystalline solids can be broadly classified as follows:

Molecular solids are those in which the binding forces between discrete molecules are due to instantaneous van der Waals interactions   in addition to dipole-dipole attractions in polar molecules. Covalent bonds exist within the molecules between the atoms.

- Generally soft
- Gas, liquid or solid at room temperature
- Low melting point (about 20 ± 200°C)
- Not usually soluble in water
- Poor conductors of electricity
- Heat insulator

Examples: molecular crystals of noble gases, halogens. Compounds such as CO_2, CH_4, H_2O, C_6H_6, $C_{12}H_{22}O_{11}$ (sucrose) and most organic compounds.

Ionic solids are those in which the binding energy is due to Coulomb electrostatic forces between positively and negatively charged ions in the crystal. That is, ionic solids are composed of **cations** and **anions**.

- Generally hard
- Solid at room temperature
- High melting point (about 800°C)
- Usually soluble in water
- Poor conductors of electricity in solid form
- Heat insulator

Examples: cations from Group I (e.g. Na^+) or Group II (e.g. Mg^{2+}) with anions O^{2-}, S^{2-}, NO_3^-, CO_3^{2-}, ClO_4^-, SO_4^{2-}, CrO_4^{2-}, PO_4^{3-}, H^+, OH^-, MnO_4^-, CN^-.

Covalent solids are those in which the binding energy is due to shared valence electrons (**covalent bonds**) between atoms in the solid.

- Generally hard
- Solid at room temperature
- Very high melting point (about 1200°C)
- Insoluble in water
- Poor conductors of electricity
- Heat insulator

Examples : Si, C (**diamond**), Ge, SiO_2 (**quartz**), BN, SiC.

Metallic solids in which valence electrons are effectively shared amongst all the atoms in the solid.

Amorphous solids have no long-range regular repeating pattern of atoms or molecules. Examples are glass and most plastics. In these materials, there is an orderly structure in the neighbourhood of any one atom, but this is not regularly repeated throughout the material.

5.12 Crystalline Lattice Structures

For geometrical reasons, there are only fourteen types of lattices that satisfy symmetry operations such as translation, rotation, reflection and inversion. Each of these fourteen lattices is called a **Bravais lattice**. There are seven convenient **crystal systems** in the set of Bravais lattices: cubic, tetragonal, orthorhombic, trigonal, monoclinic, hexagonal and triclinic.

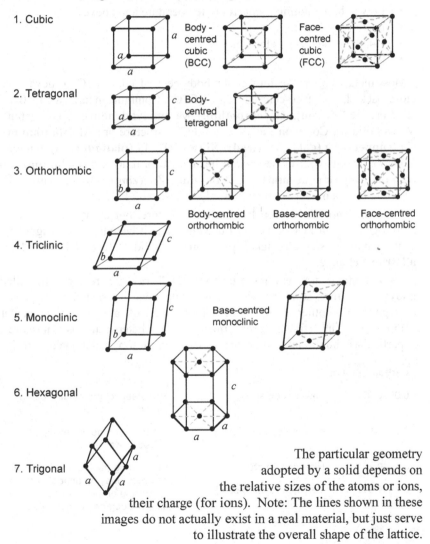

1. Cubic

Body-centred cubic (BCC) Face-centred cubic (FCC)

2. Tetragonal

Body-centred tetragonal

3. Orthorhombic

Body-centred orthorhombic   Base-centred orthorhombic   Face-centred orthorhombic

4. Triclinic

5. Monoclinic

Base-centred monoclinic

6. Hexagonal

7. Trigonal

The particular geometry adopted by a solid depends on the relative sizes of the atoms or ions, their charge (for ions). Note: The lines shown in these images do not actually exist in a real material, but just serve to illustrate the overall shape of the lattice.

5.13 Metallic Solids

Metallic solids are those in which valence electrons are effectively shared amongst all the atoms in the solid. Atoms are held in place by the **metallic bond**.

- Can be either hard or soft
- Solid at room temperature (except Hg)
- Low to high melting point (room temperature and above)
- Insoluble in water
- Good conductors of electricity
- Heat conductor

Most metals crystallise into either body-centred cubic (BCC), hexagonal close-packed, and face-centred cubic (FCC). Common metals with **BCC** structure are Fe (**iron**), Cr (**chromium**), Mo (**molybdenum**), W (**tungsten**), V (**vanadium**). Common metals with **FCC** structure are Al (**aluminium**), Cu (**copper**), Au (**gold**), Pb (**lead**), Ni (**nickel**), Pt (**platinum**), Ag (**silver**). Common metals with close-packed hexagonal form are Cd (**cadmium**), Co (**cobalt**), Mg (**magnesium**), Ti (**titanium**), Zn (**zinc**). Metals with BCC structure have a higher yield strength than those with FCC structure.

At room temperature, iron has a BCC structure, but above 910°C, iron rearranges into an FCC structure (at which time it becomes non-magnetic). This change in structure takes place in the solid state and is called an **allotropic change**.

Solid solutions are very important in **metallurgy**, where they are called **alloys**. In a true solid solution, one component is randomly dispersed throughout the other component. Other types of alloys are molecular compounds (in fixed proportions) and heterogeneous mixtures. A particularly important alloy is that of iron and carbon, which makes **steel**.

Carbon Content:

0.05–0.3%	Low-carbon steel	Mild steel, tough, ductile, easily forged and welded.
0.3–0.45%	Medium-carbon steel	Strong, hard, not so easily forged or welded.
0.45–0.75%	High-carbon steel	Very strong (high-strength steel) and hard. May be annealed for machining or heat treated for different degrees of hardness.
0.75–1.5%	Very high-carbon steel	

6. Chemical Thermodynamics

Summary

Internal energy: $\quad \Delta U = nC_p(\Delta T)$

Work: $\quad W = p\Delta V$

First Law: $\quad Q_p = \Delta U + p\Delta V$

Enthalpy: $\quad \Delta H = \Delta U + p\Delta V$

$\qquad\qquad\quad = Q_p$

Entropy: $\quad \Delta S = \dfrac{Q}{T}$

$\Delta S_{total} > 0 \quad$ irreversible process
$\Delta S_{total} = 0 \quad$ reversible process
$\Delta S_{total} < 0 \quad$ impossible process

Gibbs energy: $\quad \Delta G = \Delta H - T\Delta S$

Spontaneous reactions:

$$\Delta G_{reaction} = \Delta_f G^0{}_{products} - \Delta_f G^0{}_{reactants}$$

6.1 Molecular Energy

Molecules in a gas are capable of **independent motion**.
Consider a diatomic gas molecule:

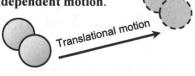

(a) the molecule itself can travel
as a whole from one place to
another

(b) the molecule can spin
around on its axis

(c) the atoms within the molecule can
vibrate backwards and forwards

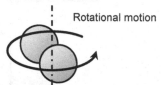

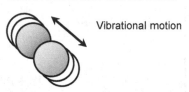

These **kinetic energy** terms represent the **internal energy** and have
significance for the interpretation of the **specific heat** capacity of a
substance. Changes in internal energy of a molecule manifest themselves
physically as changes in **temperature** of the molecule.

The total energy of the
molecule also contains some
potential energy components.
Firstly, there are **intermolecular
forces** (such as **van der Waals
forces**) that lead to the formation
of liquids, and hydrogen bonds.

Secondly, there is the
potential energy associated
with the **chemical bond**
between the atoms of the
molecule, the **bond energy**.

These potential energy terms
are associated with the *position*
of the atoms within a molecule,
and the position of molecules in
relation to other molecules, as
distinct from the *motion* of the
atoms or molecules.

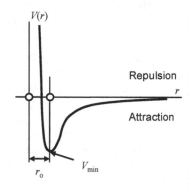

6.2 Specific Heat Capacity

The amount of energy (ΔQ) required to change the temperature of a mass of material is found to be dependent on the:

- Mass of the body (m)   kg
- Temperature increase (ΔT)   °C, K
- Nature of the material (c)   J/kg/K

$$\Delta Q = mc(\Delta T)$$ Specific heat or heat capacity

Material	c (kJ/kg/K)
water	4.186
steel	0.45
cast iron	0.54
aluminium	0.92

The **specific heat** is the amount of heat required to change the temperature of 1 kg of material by 1°C

Molar specific heat, or **molar heat capacity** (C), is the amount of heat required to raise the temperature of 1 mole of the substance by 1°C.

$$\Delta Q = nC(\Delta T)$$

Experiments show that when a gas is heated at constant volume, the **molar heat capacity** C_v is always less than that if the gas is heated at constant pressure C_p. This is because for a constant volume process, there is no opportunity for there to be mechanical work done on or by the system.

For a constant pressure process, for a given temperature rise ΔT, there will always be a volume change ΔV. Therefore the energy into the system has to both raise the temperature *and* do work, thus C_p is always greater than C_v.

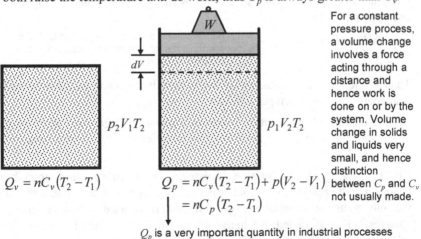

$$Q_v = nC_v(T_2 - T_1)$$

$$Q_p = nC_v(T_2 - T_1) + p(V_2 - V_1)$$

$$= nC_p(T_2 - T_1)$$

For a constant pressure process, a volume change involves a force acting through a distance and hence work is done on or by the system. Volume change in solids and liquids very small, and hence distinction between C_p and C_v not usually made.

Q_p is a very important quantity in industrial processes and is given the name **enthalpy**.

6.3 Enthalpy

Consider a beaker of solution which is heated from an external source. That is, heat energy is transferred into the system across the boundary.

The temperature of the solution increases by an amount dependent on the **molar heat capacity** C_V of the solution. The change in **internal energy** of the system will be:

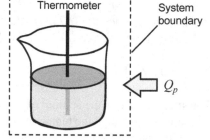

$$\Delta U = nC_v(\Delta T)$$

The solution may also expand by an amount ΔV and perform mechanical **work** (e.g. by a piston that fits into the top of the beaker and floats on the surface of the solution):

$$W = p\Delta V$$

The relationship between the heat entering the system, the work done by the system and the rise in internal energy is expressed by the **first law of thermodynamics:**

$$Q_p = \Delta U + p\Delta V$$

Q_p is positive when heat enters the system. W is positive when work is done by the system on the surroundings. ΔU is positive when ΔT is positive.

If heat energy Q_p is added to the system, then some of this heat goes into mechanical work and some into internal energy. The sum of the change in internal energy and the work done on or by the system is called the change in **enthalpy** ΔH of the system:

$$\Delta H = \Delta U + p\Delta V$$
$$= Q_p$$

Note: $p\Delta V$ is usually small in comparison to ΔU in experiments in the laboratory whereupon ΔH is approximately equal to ΔU.

That is, the change in enthalpy of the system is the same as the amount of heat energy that has been put into it.

If the heating of the solution involves a **phase change**, then some of the heat energy input would be consumed by latent heat which would have to be accounted for.

If the heating of the solution initiates a **chemical reaction**, then heat energy may be released (e.g. as happens during combustion) or absorbed by the solution as the potential energy of the **chemical bonds** (chemical energy) changes. This is called the **heat of reaction**. The heat of reaction is a release, or absorption of potential energy of the chemical bonds.

6.4 Heat of Reaction

Consider the reaction between two chemical species in an aqueous solution in a beaker which is placed in a temperature controlled water bath so that the beaker is maintained at a constant temperature. That is, heat flow can occur across the system boundary but the solution is held **isothermal**.

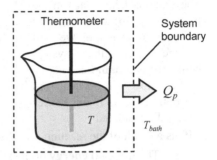

If, as a result of the reaction, heat energy is released, then this heat must pass out through the system boundary. The surrounding water bath ensures that the heat energy is taken away fast enough to prevent any temperature rise of the solution.

- The **internal energy** of the system remains the same after the reaction as indicated by no rise in temperature.
- **Energy** has come from the **heat of reaction** and been transferred into the water bath.
- If there has been a volume change, then mechanical **work** also would have been performed on (+ve) or by (-ve) the system.

In this example, there must have been a net release of bond energy as a result of the chemical reaction. Some of this energy has been carried away across the boundary by the water bath, and some (a small amount in this case) has gone into mechanical work against the pressure of the atmosphere. The change in **enthalpy** of the system is:

$$\Delta H = \Delta U + p\Delta V$$

We can assume that the mechanical work involved in this example is small compared to the heat lost by the system to the surroundings. Thus, $\Delta H = Q_p$. In this example, heat Q_p has passed out of the system and so the change in enthalpy is a decrease. When heat is passed out to the surroundings, the reaction is said to be **exothermic**. If the reaction involved an absorption of heat from the surroundings, then the change in enthalpy would be positive, and the reaction is called **endothermic**.

The heat of reaction represents the change in enthalpy in this isothermal system. The question now is, if there is a chemical reaction, *and* the temperature changes, how is the change in enthalpy calculated?

6.5 Heat of Reaction

Consider the reaction between two chemical species in an aqueous solution in an insulated beaker. A **thermometer** measures the temperature change during the reaction.

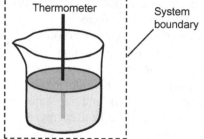

The reaction takes place inside the boundary of a system. For the present example, we will assume that the system boundary is **adiabatic**. That is, no heat flow will occur across the boundary, as if the beaker were fully insulated.

If, as a result of the reaction, the temperature of the solution rises, then:

- The **internal energy** of the system is higher after the reaction, as indicated by a rise in temperature.

- **Energy** has come from the **heat of reaction** and been transferred into internal energy of motion of the molecules in the solution.

- If there has been a volume change, then mechanical **work** also would have been performed on (+ve) or by (-ve) the system.

Since no external heat energy was involved (no heat flow across the system boundary), the heat energy needed to raise the temperature of the solution must have come from the net sum of energies from the breakage and formation of the **chemical bonds** involved in the reaction. In this example, there must have been a net release of bond energy as a result of the reaction. Some of this energy has gone into internal energy of motion of the molecules, while some (a small amount in this case) has gone into mechanical work against the pressure of the atmosphere. In this case, $\Delta H = \Delta U + p\Delta V$. If the work done is negligible, then $\Delta H \approx \Delta U$. In this case, the change in enthalpy is equal to the **heat of reaction**. Although no heat has crossed the system boundary, heat has appeared as a result of the chemical reaction – just as if it had come from outside the system. That is, in this *adiabatic* example, heat Q_p has entered the system from the chemical reaction and $\Delta H \approx \Delta U$.

6.6 Enthalpy of Formation

A chemical reaction is the process of breakage and formation of chemical bonds. Energy is absorbed when a bond is broken, and energy is released when a bond is formed. When a chemical reaction occurs, heat may be either given off, or absorbed to or from the surroundings. That is, the change in enthalpy may be either positive (**endothermic**) or negative (**exothermic**).

Liquid water molecules are made by the reaction between H_2 and O_2 gas molecules:

$$H_2(g) + \frac{1}{2}O_2(g) \rightarrow H_2O(aq)$$

The equation is written in terms of the formation of 1 mole of product. Under standard conditions (1 atm pressure and 25°C) it is found that 286 kJ of heat energy is produced. That is, the change in enthalpy of the system is $\Delta H = -286$ kJ. This is the **heat of reaction**.

When the reactants are at standard conditions, and the products are also kept at standard conditions (e.g. by use of a water bath to maintain temperature), the heat of reaction is referred to as the **heat of formation**, or **enthalpy of formation**. Standard heats of formation are determined by experiment.

Name	Molecular Formula	Heat of Formation ΔH, kJ/mol
Methane	CH_4	-75
Carbon dioxide	CO_2	-394
Ammonia	NH_3	-46
Acetylene	C_2H_2	+227
Ethylene	C_2H_4	+227
Ethane	C_2H_6	-85
Water	H_2O	-286
Hydrochloric acid	HCl	-92
Sodium hydroxide	NaOH	-427
Aluminium oxide	Al_2O_3	-1670
Carbon monoxide	CO	-110

When the reaction proceeds in reverse, the sign of the heat of formation is reversed.

The reference point for heats of formation are elements at standard conditions. For example the heat of formation of $O_{2(g)}$ is set to zero.

Heats of any reaction can be obtained by addition of heats of formations of known compounds, or **Hess' law**. The change in enthalpy for the combustion of ethane is:

$$C_2H_6 + \frac{7}{2}O_2 \rightarrow 2CO_2 + 3H_2O$$

$$\left(C_2H_6(g) \rightarrow 2C(s) + 3H_2(g)\right) \quad \Delta H = +85$$

$$2\left(C + O_2 \rightarrow CO_2\right) \qquad \Delta H = 2(-394)$$

$$3\left(H_2 + \frac{1}{2}O_2 \rightarrow H_2O\right) \qquad \Delta H = 3(-286)$$

$$\Delta H = 85 - 2(394) - 3(286) = -1561 \text{ kJ/mol}$$

Note that the sign of ΔH for C_2H_6 has been reversed in this calculation since the reaction splits this compound rather than forms it. ΔH for the combustion of ethane is called the **calorific value**.

6.7 Entropy

Consider a beaker of ice which is heated from an external source. The ice remains at 0°C (273K) and melts. The amount of energy Q_p needed is determined by the **latent heat of fusion** of ice.

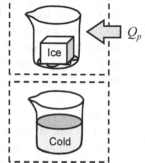

The water molecules have a greater amount of disorder after melting compared to that before heating due to their increased velocity. A quantitative measure of disorder is **entropy**.

Absolute values of entropy S can be calculated from thermodynamic theory or measured in the laboratory. Of more interest to us in chemistry are changes in entropy as a result of physical changes of state, and chemical reactions. When a system undergoes a process at temperature T involving heat flow, the *change* in entropy of the **system** is:

$$\Delta S = \frac{Q}{T}$$

When a system absorbs heat, Q is positive and so the entropy increases, $\Delta S > 0$. When a system rejects heat, Q is negative and the entropy decreases, $\Delta S < 0$.

If the process occurs at a low temperature, then the change in entropy is greater than if the same process occurs at a higher temperature. At high temperatures, the entropy is already high and so the proportional change in entropy for a given heat flow Q is less compared to that which would occur at low temperatures.

For 1 mole ice:

$$\Delta S = \frac{mL}{T}$$

$$= \frac{0.018(335000)}{273}$$

$$= 22.1 \text{ J/K}$$

When a system undergoes a process and interacts with its surroundings (i.e. a non-isolated system), the total **entropy change** (of the universe) is:

$$\Delta S_{total} = \Delta S + \Delta S_{surroundings}.$$

The sign of ΔS_{total} signifies the presence of the following types of processes:

$\Delta S_{total} > 0$ irreversible process
$\Delta S_{total} = 0$ reversible process
$\Delta S_{total} < 0$ impossible process

Entropy is not energy. There is no "law of conservation of entropy". Irreversible processes *create* entropy.

When a phase change or a chemical reaction involves heat flow to or from the surroundings, entropy must be considered. The **total entropy** of the system and the surroundings cannot decrease. If the total entropy change is zero, then the process is reversible and so cannot be **spontaneous** in any one direction. For a reaction to proceed spontaneously from reactants to products, the total entropy ΔS_{total} must increase.

6.8 Entropy Calculations

Changes in entropy of a system at constant temperature are expressed as:

$$\Delta S = \frac{Q}{T}$$

When the temperature is not constant, we add up incremental changes in entropy for incremental transfers of heat:

$$dQ = mcdT$$

$$\Delta S = \int_{Q_1}^{Q_2} \frac{1}{T} dQ$$

$$= \int_{T_1}^{T_2} \frac{1}{T} mc\, dT$$

$$= mc \ln \frac{T_2}{T_1}$$

The **standard molar entropy** S° is the entropy content of one mole of substance under standard conditions of 1 atm pressure and 25°C. Standard entropies are absolute values and represent the change in entropy to bring the substance from 0K to 273K in its standard state. Standard entropies are always positive. The zero point for entropy is 0K. This statement of zero point for entropy is called the **third law of thermodynamics**.

For a chemical reaction or a physical process, the change in entropy can be calculated from:

$$\Delta S_{reaction} = S^\circ_{products} - S^\circ_{reactants}$$

For example:

$$H_2(g) + \frac{1}{2}O_2(g) \rightarrow H_2O(g)$$

$$\Delta S^\circ = 188 - (131 + 0.5 \times 205)$$

$$= -45.5 \text{ J/K / mol}$$

Element	S°. (J/K/mol)
$H_2(g)$	131
$O_2(g)$	205
$N_2(g)$	192
$Si(s)$	19
$C(s)$	6
$Fe(s)$	27

Compound	S° (J/K/mol)
$NH_3(g)$	192.8
$H_2O(l)$	70.0
$H_2O(g)$	188.8
$CO_2(g)$	213.8
$CO(g)$	197.7
$CH_4(g)$	186.2

Note that in this example, the change in entropy for this system is < 0 at the expense of an increase in entropy elsewhere in the universe. Hydrogen and oxygen will, under certain conditions, form water since it is energetically favourable to do so.

6.9 Gibbs Energy

It is a natural tendency of systems to attain the lowest possible potential energy, such as a ball rolling down a hill, electrons moving in an electric field, and so on. In an **exothermic** process, heat leaves a system ($\Delta H < 0$) and we might expect that such processes are always spontaneous and that **endothermic** processes are not. However, this is not observed. For example, when water evaporates spontaneously at room temperature, $\Delta H > 0$ and the water vapour has a higher energy than the liquid.

$$H_2(g) + \frac{1}{2}O_2(g) \rightarrow H_2O(aq) \quad \Delta H = -286 \text{ kJ/mol}$$

$$H_2(g) + \frac{1}{2}O_2(g) \rightarrow H_2O(g) \quad \Delta H = -242 \text{ kJ/mol}$$

$$H_2O(aq) \rightarrow H_2O(g) \quad\quad\quad \Delta H = 286 + -242 = +44 \text{ kJ/mol}$$

$+\Delta H$ signifies energy of products is greater than reactants (endothermic)

Whether or not a phase change or a chemical reaction proceeds spontaneously cannot be judged on the basis of energy alone. The change in entropy ΔS_{total} of both the system (the reaction) and the surroundings must also be considered. Whether or not the change in entropy or enthalpy is the more important factor in any particular process depends on the absolute temperature T.

Now, the total energy of a system is given by the **enthalpy** and this consists of the internal energy and the external work. Changes in total energy are expressed as: $\Delta H = \Delta U + p\Delta V$.

Part of the total energy H is unavailable for conversion into work. This part is connected with changes in **entropy**. Since the units of entropy are J/K, and the determining factor that quantifies the contribution of entropy in a particular process is T, then the unavailable energy can be calculated from the product TS. Thus, the total energy of a *system* can be written as:

$$H = G + TS$$

where G represents that part of the total energy that is freely available for conversion into work. This freely available energy is called the **Gibbs energy**. As the temperature increases, the amount of Gibbs energy decreases and the unavailable term TS increases. Changes in the amount of Gibbs energy in a constant temperature process are thus: $\Delta G = \Delta H - T\Delta S$.

The concept of Gibbs energy allows us to predict the direction of spontaneous chemical reactions by considering both energy and entropy. A spontaneous chemical reaction will proceed only when there is a decrease in available Gibbs energy: $\Delta G < 0$.

6.10 Spontaneous Processes

The concept of Gibbs free energy embodies both energy and entropy considerations in determining in which direction a chemical reaction proceeds. A decrease in Gibbs energy can arise due to a decrease in **enthalpy** or an increase in the product TS. In a particular chemical reaction, ΔH and ΔTS may have the same or opposite signs. For a constant pressure process, at constant temperature:

$$\Delta G = \Delta H - T\Delta S.$$

- If a reaction is exothermic $\Delta H < 0$ and the entropy change $\Delta S > 0$, then the free energy change $\Delta G < 0$ and the process is **spontaneous**.
- If a reaction is endothermic $\Delta H > 0$ and the entropy change $\Delta S < 0$, then the free energy change $\Delta G > 0$ and the process is not spontaneous.
- If the enthalpy change $\Delta H > 0$ and the entropy change $\Delta S > 0$, then the process is spontaneous at high temperature. When $\Delta H < 0$ and $\Delta S < 0$, the process is spontaneous at low temperature.

It is important to know that ΔS in the above equation is the entropy change of the *system*, not the total entropy change (of the universe). This enables us to conveniently predict and quantify the direction of a process based upon attributes of the system only.

In chemical reactions, the change in Gibbs energy depends on the physical states of the reactants and products, the temperature and pressure, and the concentration of the reactants and products. When reactants and products are at standard conditions (1 atm and 25°C), the free energy change is referred to as the **standard Gibbs energy change** ΔG°.

The standard Gibbs energy change of the formation of a compound from its elements is called the **standard Gibbs energy of formation** $\Delta_f G^\circ$. Elements at their standard states have zero standard Gibbs energy of formation.

For a process or a chemical reaction, the change in Gibbs energy can be calculated from:

$$\Delta G_{reaction} = \Delta_f G^\circ{}_{products} - \Delta_f G^\circ{}_{reactants}$$

Compound	$\Delta_f G^\circ$. (kJ/mol)
$NH_3(g)$	−16.6
$H_2O(l)$	−237.2
$H_2O(g)$	−228.6
$CO_2(g)$	−394.3
$CO(g)$	−137.3
$CH_4(g)$	−50.8

6.11 Melting of Ice

Consider a mole of ice initially at 0°C. Now, let the temperature of the surroundings be 283K (+10°C).

For the system: $\Delta H = mL + mc\Delta T$

$$= 0.018(335000) + 0.018(4186)(10)$$

For the surroundings:

$$= 6780 \text{ J} \qquad \text{(endothermic)}$$

$$\Delta S_{\text{surroundings}} = \frac{-6780}{283} = -23.9 \text{ J/K}$$

For the system: $dQ = mL + mcdT$

$$\Delta S = \frac{mL}{T_1} + \int_{Q_1}^{Q_2} \frac{1}{T} dQ$$

$$= \frac{mL}{T_1} + \int_{T_1}^{T_2} \frac{1}{T} mc \, dT$$

$$= \frac{0.018(335000)}{273} + 0.018(4186)\ln\frac{283}{273} = 24.8 \text{ J/K}$$

Total change in entropy:

$$\Delta S_{\text{total}} = \Delta S + \Delta S_{\text{surroundings}}$$

$$= 24.8 - 23.9$$

$$= +0.9 \text{ J/K}$$

This is an **irreversible process**. Even though the change in entropy is < 0 for the *surroundings*, the change in entropy of the system is greater, and so the change in **total entropy** is > 0.

$$\Delta G = \Delta H - T\Delta S$$

$$= 6780 - 283(24.8)$$

$$= 6780 - 7018$$

$$= -238 \text{ J}$$

The **Gibbs free energy** change is < 0 so this process is spontaneous.

Data:
c_{water} = 4186 J/kg/K
c_{Al} = 920 J/kg/K
L_v = 22.57 × 10^5 J/kg
L_f = 3.35 × 10^5 J/kg
ρ_{water} = 1000 kg/m^3
c_{ice} = 2110 J/kg/K

6.12 Freezing of Water

Consider a mole of water, initially at 0°C. Now, let the temperature of the surroundings be 263K (−10°C).

For the system: $\Delta H = -mL + -mc\Delta T$

$$= -0.018(335000) + -0.018(2110)(10)$$

$$= -6.03 - 0.38$$

For the surroundings: $= -6410$ J   (exothermic)

$$\Delta S_{surroundings} = \frac{6410}{263} = 24.4 \text{ J/K}$$

For the system: $dQ = -mL + mcdT$

$$\Delta S = \frac{-mL}{T_1} + \int_{Q_1}^{Q_2} \frac{1}{T} dQ$$

$$= \frac{-mL}{T_1} + \int_{T_1}^{T_2} \frac{1}{T} mc \, dT$$

$$= \frac{-0.018(335000)}{273} + 0.018(2110)\ln\frac{263}{273} = -23.52 \text{ J/K}$$

Total change in entropy:

$$\Delta S_{total} = \Delta S + \Delta S_{surroundings}$$

$$= -23.52 + 24.4$$

$$= +0.88 \text{ J/K}$$

$$> 0$$

This is an **irreversible process**. Even though the change in entropy is < 0 for the *system*, the change in entropy of the surroundings is greater, and so the change in **total entropy** is > 0.

$$\Delta G = \Delta H - T\Delta S$$

$$= -6410 - 263(-23.5)$$

$$= -6410 + 6180$$

$$= -229 \text{ J}$$

This is a spontaneous process since the freezing of ice results in a decrease in **Gibbs free energy**.

Data:
$c_{water} = 4186$ J/kg/K
$c_{Al} = 920$ J/kg/K
$L_v = 22.57 \times 10^5$ J/kg
$L_f = 3.35 \times 10^5$ J/kg
$\rho_{water} = 1000$ kg/m^3
$c_{ice} = 2110$ J/kg/K

6.13 Ice/Water Equilibrium

Consider the conversion of one mole of water to ice at 273K:

For the system: $\Delta H = -mL$
$$= -0.018(335000)$$
$$= -6030 \text{ J} \quad \text{(endothermic)}$$

For the surroundings:
$$\Delta S_{\text{surroundings}} = \frac{6030}{273} = 22.08 \text{ J/K}$$

For the system: $dQ = -mL$

$$\Delta S = -\frac{mL}{T}$$
$$= \frac{-0.018(335000)}{273}$$
$$= -22.08 \text{ J/K}$$

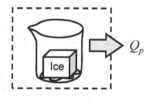

Total change in entropy:

$$\Delta S_{\text{total}} = \Delta S + \Delta S_{\text{surroundings}}$$
$$= -22.08 + 22.08$$
$$= 0 \text{ J/K}$$

The total change of entropy is zero. This is a **reversible process**. Extraction of Q_p from the melted ice would cause it to re-freeze.

$$\Delta G = \Delta H - T\Delta S$$
$$= -6030 - 273(-22.08)$$
$$= 0 \text{ J}$$

This process is at **phase equilibrium**.

Data:
c_{water} = 4186 J/kg/K
c_{Al} = 920 J/kg/K
L_v = 22.57 × 10^5 J/kg
L_f = 3.34 × 10^5 J/kg
ρ_{water} = 1000 kg/m^3
c_{ice} = 2110 J/kg/K

6.14 Chemical Equilibrium

A chemical reaction involves reactants and products. Whether or not a chemical reaction proceeds spontaneously depends on the sign of the change in free energy.

$$\Delta G_{\text{reaction}} = \Delta_f G^0_{\text{products}} - \Delta_f G^0_{\text{reactants}}$$

$$\text{Reactants} \rightarrow \text{Products}$$

If a chemical reaction proceeds from reactants to products spontaneously, we may at first expect the process to continue until the reactants are all converted to product. This is not observed. Usually, reactions proceed until there is **chemical equilibrium**:

$$\text{Reactants} \leftrightarrow \text{Products}$$

At equilibrium, the rate of forward reaction = rate of reverse reaction.

In chemical reactions, the **free energy** change depends on the temperature, physical state, pressure, and concentrations of the reactants and the products. That is, as the reaction proceeds, the reactants are consumed and the products are created – and the free energy change changes as the reaction proceeds in time. When the free energy change becomes equal to zero, the rate of forward reaction has become equal to the rate of the reverse reaction. This usually happens (for spontaneous reactions) when the concentration of the products is somewhat greater than those of the reactants – but it is important to note that this does not necessarily mean that all the products are used up. At **chemical equilibrium**, the forward and reverse reactions are both happening at the same rate.

Precisely what is meant by **rate of reaction** depends upon the context. Rates of reaction may be expressed as the change in concentration of one of the reactants per unit time where concentration is expressed in terms of moles per unit volume.

State	Concentration units
Gases	Partial pressure units
Liquids	Molarity (moles per litre)
Solids	Moles or mass units

The appropriate time period may be expressed in milliseconds, seconds, days or years depending on the reaction.

6.15 Statistical entropy

The change in entropy of a system can be calculated from macroscopic quantities, Q and T. Molecules of gas in a volume V at temperature T have a distribution of velocities, and hence, kinetic energies. The distribution of velocities over the total number of molecules was determined by Boltzmann:

The number of molecules N with an energy E was computed by Boltzmann using the **Maxwell velocity distribution** function:

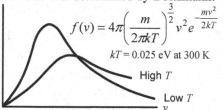

$$f(v) = 4\pi \left(\frac{m}{2\pi kT} \right)^{\frac{3}{2}} v^2 e^{-\frac{mv^2}{2kT}}$$

$kT = 0.025$ eV at 300 K

High T

Low T

$$N = Ce^{\frac{-E}{kT}} \quad \begin{array}{l} \text{Maxwell-Boltzmann} \\ \text{energy distribution} \end{array}$$

Consider a partitioned volume V containing N molecules. There is a discrete number of energies possible, and there is a finite number of ways that the molecules present can have a distribution of energies that matches the macroscopic properties of P, T and V. Each of these possibilities is called a **microstate**.

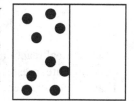

Now, one such possible state is if all the energy of the system were concentrated in one molecule and the others had zero energy. We know that this is very unlikely to ever happen, just as we know that the likelihood of throwing zero heads and 100 tails in 100 tosses of a coin is unlikely. The most likely scenario for our N molecules is that the energies will follow the Maxwell-Boltzmann distribution.

In mathematical symbols, the **statistical weight function** of the system $\Omega(E,V,N)$ is the number of possible microstates that are consistent with the observed **macrostate**. The lower the number of possible microstates, the more accurately we know information about the molecules. The entropy of the system is: $S = k \ln \Omega$.

If there were only one possible microstate, then S would be zero. As the number of possible microstates increases, the entropy becomes larger. When the partition is removed, the molecules rush to fill the entire volume. It is exceedingly unlikely that at any one instant, all the molecules will be located over on one half of the now larger volume. The number of possible microstates has increased. However, one might ask what if the container had only two molecules in it? In this case, it is likely that at any one time, the two molecules *will* find themselves over to one side of the container. The number of possible microstates, for the two-molecule scenario, is very small, and so is the entropy.

7. Rates of Reaction

Summary

Rates of reactions:
- The nature of the reactants
- The concentration of the reactants
- The temperature
- The presence of a catalyst

Rate law (concentrations):

$$a\text{A} + b\text{B} \rightarrow \text{P}$$

$$\frac{d[\text{P}]}{dt} = k[\text{A}]^n[\text{B}]^m$$

Rate law (temperature):

$$k = Ae^{\frac{-E}{RT}} \quad \text{Arrhenius equation}$$

Catalyst:
- Has no effect on the stoichiometry of the overall reaction.
- Does not affect the equilibrium position of the reaction.
- Affects the reaction mechanism so as to lower the activation energy, thus increasing the number of molecules that can make a productive collision.

7.1 Rates of Reaction

Consider the reaction between hydrogen gas and oxygen gas to produce liquid water:

$$H_2(g) + \frac{1}{2}O_2(g) \rightarrow H_2O(aq) \quad \Delta H = -286 \text{ kJ/mol}$$

$$H_2(g) + \frac{1}{2}O_2(g) \rightarrow H_2O(aq)$$

$$\Delta S^\circ = 70 - (131 + 0.5 * 205)$$
$$= -163.3 \text{ J/K/mol}$$

$$\Delta G = \Delta H - T\Delta S$$
$$= -286000 - 273(-163.3)$$
$$= -286000 + 44580.9$$
$$\approx -240 \text{ kJ/mol}$$

The indications are that this reaction should proceed spontaneously when these two gases are mixed. While this might be so, experience indicates that this reaction does not proceed at all unless there is a spark or source of ignition present – in which case the reaction proceeds with explosive violence.

Clearly then, the **rates of reaction** are an important part of chemical reactions. While thermodynamics might provide information about the direction in which a reaction might proceed on its way to chemical equilibrium, more information is needed to determine at what rate the products are produced.

Associated with the rate of reaction is the **reaction mechanism**. The production of products and the consumption of reactants, while observable, only represents the net reaction in general. Chemical reactions may proceed in several steps. Intermediate species may form and be consumed, and these intermediate steps often influence the overall rate of the reaction as a whole.

The most important factors that govern the rate of a reaction are
- The nature of the reactants
- The concentration of the reactants
- The temperature
- The presence of a catalyst

The rate of reaction is the rate of change of **concentration** of either the reactants or the products.

7.2 Collision Theory

Chemical reactions involve the breaking and formation of bonds between atoms, ions or molecules. Elements or compounds that comprise the reactants form new compounds as products. The relationship between energy and entropy – the **Gibbs free energy** – determines which direction a chemical reaction will proceed. The question now is to ask just how a chemical reaction takes place.

The notion of chemical reactions only taking place between atoms or molecules when they contact each other in a certain way is not so obvious as it might seem and was only proposed in the early 1900s.

Central to the answer of this question is the fact that atoms, ions and molecules are in constant motion – especially when they are in liquid and gaseous form. Because they are in motion, they undergo collisions with other atoms and molecules.

When atoms, ions or molecules collide, a chemical reaction may occur. That is, chemical reactions occur as a result of collisions between molecules. This is known as **collision theory**.

No reaction

Reaction

During the collision, bonds are broken and formed as atoms and electrons are rearranged into new compounds. Several variables serve to determine how fast reactants are converted into products:

- Particles (either atoms, ions or molecules) have to collide in a particular way. That is, the orientation of the colliding particles has to be a certain way in order that the collision be effective at breaking bonds and forming new ones.

- The particles have to possess enough kinetic energy to allow them to interact to the extent that bonds may be broken and allow new ones to be formed.

- The number of collisions per unit time determines how fast the reaction proceeds and this is related to the concentration of the reactants.

The nature of the reactants, the temperature, and concentration and action of **catalysts** can be all explained by collision theory. Not all collisions result in a reaction. Only a small proportion are **productive collisions**.

7.3 Reaction Mechanism

We have agreed that for a chemical reaction to proceed, atoms, ions, or molecules must collide. Further, they must not just touch or brush past one another. They must collide with a certain amount of kinetic energy, and be in the right orientation with respect to one another in order for bonds to be broken and allow new ones to form. That is, there has to be a **productive collision**.

Consider the rusting of iron. Simply put, the reaction is:

$$4Fe(s) + 3O_2(g) \rightarrow 2Fe_2O_3(s).$$

This is a reasonably slow reaction. However, if one were to just take collision theory at face value, it would be luck indeed if four atoms of iron were to collide with three molecules of oxygen at the same time to form two molecules of iron oxide. In reality, reactions usually proceed as a series of smaller steps. The series of smaller steps is called the **reaction mechanism**.

$$Fe^{2+} + 2OH^- \rightarrow Fe(OH)_2$$

$$4Fe(OH)_2 + O_2 + 2H_2O \rightarrow Fe(OH)_3(s)$$

$$Fe(OH)_3(s) \rightarrow Fe_2O_3, H_2O + 2H_2O$$

The chances of a collision occurring between two atoms, ions or molecules is significantly greater than the chances of a collision between three or four of the required chemical species at the same time. Most reaction steps in chemical reactions involve two atoms or molecules. The slowest step in the reaction mechanism is the **rate-determining step**. The nature of the reaction steps cannot be deduced from the net reaction. The reaction steps can only be found by experiment.

Identification of the rate-determining step is important for establishing a quantitative rate law for the reaction as a whole. Often, the rates of reaction depend on the concentration of the reactants, but it is the concentration of the reactants within the rate-determining step that is they key issue. During the **elementary reactions**, or the **reaction steps**, intermediate chemical species may be formed which are used up in a later step and do not appear in the net equation.

Some reaction mechanisms are noted for their explosive properties. In these cases, the steps become a self-sustaining **chain reaction** (such as in **combustion**).

7.4 Activation Energy

A reaction step usually involves a two-particle collision (either atoms, ions or molecules). As two molecules approach each other, the outer electrons of one molecule will begin to feel the electrostatic repulsion of the other molecule. The kinetic energy of motion will be converted into electrical potential energy. The potential energy rises as the molecules come closer and closer together.

When all the kinetic energy has been expended, the molecules are in very close proximity and are in a somewhat unstable state. The molecules at this point form what is called an **activated complex**. The activated complex is a conglomeration of the participating atoms. At this point, the activated complex may split into the product molecules (the forward reaction) or bounce backwards and reform the original molecules (the reverse reaction).

The colliding molecules require sufficient **kinetic energy** to overcome mutual repulsion and allow the collision and formation of the activated complex. This is called the **activation energy** E_{act}. The activation energy depends on the nature of the reactants.

When a spark is introduced into a mixture of H_2 and O_2, the energy of the spark overcomes the activation energy and allows the reaction to proceed. The heat of the reaction furnishes the required activation energy for further reactions. That is, the H_2 gas undergoes **ignition**.

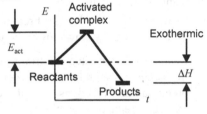

The activation energy is the difference in energy between the activated complex and the average energy of the reactants.

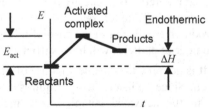

If a particular reaction has a large activation energy, then only a small proportion of collisions may result in enough kinetic energy to overcome it, and the reaction will be slow. If a particular reaction has a lower activation energy, then a greater proportion of collisions will have enough kinetic energy to overcome it and the reaction proceeds at a faster rate.

7.5 Nature of Reactants

The rate of a chemical reaction, that is, the rate of change of concentration of either the reactants or the products, depends on the nature of the reactants. When the permanganate ion is reduced by an Fe^{2+} ion in an acidic solution, the reaction is very fast and is limited by the time it takes to mix the solutions:

$$5Fe^{2+}(aq) + MnO_4^-(aq) + 8H^+(aq) \rightarrow 5Fe^{3+}(aq) + Mn^{2+}(aq) + 4H_2O \quad \text{fast}$$

But when the permanganate ion is reduced by oxalic acidic $H_2C_2O_4$, the reaction is fairly slow, taking several minutes.

$$5C_2O_4^{2-}(aq) + 2MnO_4^-(aq) + 16H^+(aq) \rightarrow 10CO_2(g) + 2Mn^{2+}(aq) + 8H_2O$$
$$\text{slow}$$

The main difference between the two reactions is the reducing agent. It could be argued that in the case of $H_2C_2O_4$, many more bonds must be broken compared to the case of the Fe^{2+} solution and this might influence the rate of the reaction. That is, the reaction mechanism for the second reaction may be quite different in character than that of the first reaction.

Experiments indicate that:

- Reactions that involve the breaking and formation of **covalent bonds** are usually slow at room temperature.

- Reactions that involve electron transfer, such as **ionic reactions** in aqueous solutions, are generally fast. Examples are precipitation reactions, **acid** on **base** reactions, most oxidation/reduction reactions.

Since the **activation energy** depends on the chemical species involved, then so does the rate of reaction. There is more chance of a successful or productive collision in the case of a low activation energy because a greater number of particles will have sufficient kinetic energy to overcome it.

It is not only the chemical nature of the reactants, but the physical nature as well. Since collision theory requires atoms or molecules to make contact to form the activated complex, any mechanism that affects the degree of contact between atoms or molecules will affect the rate of reaction. In homogenous systems, such as aqueous solutions, there is intimate contact, whereas in inhomogeneous systems, such as solid/liquids, or solid/gas, reactions only take place at the surface of contact of the atoms or molecules. Increasing the surface area by mixing or grinding increases the rate of reaction. A common example is that sawdust is almost explosive in air, while a block of wood burns relatively slowly in comparison.

7.6 Concentration

One of the most important effects on the rate of reaction is the **concentration** of one or more of the reactants. Generally speaking, increasing the concentration of a reactant increases the rate of the reaction (but not always). It is easy to accept that the chances of a successful, or productive, collision between two reacting particles is increased if there are more of them, per unit volume, involved in the process. Consider a reaction:

$$A + B \rightarrow AB.$$

If we only had one molecule of A and one of B, then we may have to wait some time before they encounter each other and collide in a way so as to produce the reaction. The collision rate, and hence the reaction rate, would be low.

If we now increase the concentration of A, but putting more molecules of A in the reaction chamber, then there would be an increase in the chances of a successful collision – the reaction rate would increase. It is possible that doubling the concentration of one or other of the reactants would double the reaction rate.

The concentration of a species in moles per litre can be conveniently written using [ ] notation. Thus, the reaction rate is indicated by the rate of change of appearance of the product, or the rate of change of concentration of AB.

Reaction rate is expressed as moles/litre/second

$$\frac{d[AB]}{dt} = k[A][B]$$

This equation is called a **rate law**. k is the constant of proportionality, or the **rate constant**. It should be noted that the rate law actually observed for a reaction cannot be necessarily deduced from the balanced net equation. This is because one of the intermediate reaction steps may be the rate-determining reaction, in which case the rate law for that step would be required. Unless more of the reactants are continually added, the rate of reaction will generally decrease with time as the reactants are consumed (and their concentrations decrease).

7.7 Rate Law

Consider the reaction: $A + A \rightarrow A_2$

$$2A \rightarrow A_2 \quad \text{Balanced equation}$$

The **rate law** would be written:

$$\frac{d[A_2]}{dt} = k[A][A]$$

$$= k[A]^2 \quad \text{Rate law}$$

Note that the exponent of [A] in the rate law is the same as the coefficient of A in the reactants in the balanced equation. This is a second-order reaction.

In general, for an equation of the form: $aA + bB \rightarrow P$ the rate law is written as:

$$\frac{d[P]}{dt} = k[A]^n[B]^m.$$

For a single-step reaction, the exponents n and m are equivalent to a and b. However, a chemical reaction nearly always proceeds via a **reaction mechanism**, that is, a series of reaction steps. Thus, in most cases, there is no connection between the exponents n, m and the coefficients a, b for the net reaction. In general, the rate law for the net reaction has to be determined by experiment.

The exponents n and m indicate the **order of the reaction** with respect to that reactant. For example, when $n = 2$ and $m = 1$, we say that the reaction is second order with respect to A and first order with respect to B. The overall order of the reaction is the sum $m+n$. When n or m equals zero, then the rate of reaction is independent of the concentration of the reactant and is a **zero-order reaction** with respect to that reactant.

The **rate constant**, k, is specific to the particular reaction and depends on the experimental conditions. The units of k depend on the order of the reaction. A high value of k indicates fast reaction. For first-order reactions, if $k > 10 \text{ sec}^{-1}$, then the reaction is considered instantaneous.

The **rate law** shows how the instantaneous rate of a reaction changes with the concentration of the reactants, as long as other factors such as temperature, the nature of the reactants and the presence of catalysts remain constant.

The rate of a reaction with respect to either reactants or products is expressed in terms of the **rate constant** k and the **order** of the reaction.

7.8 Rates of Reactions

There are several methods of expressing the rates of a reaction:

Average rate of disappearance of a reactant R: $\quad -\dfrac{[R]_2 - [R]_1}{t_2 - t_1}$

Average rate of appearance of a product P: $\quad \dfrac{[P]_2 - [P]_1}{t_2 - t_1}$

Instantaneous rate of disappearance of a reactant: $\quad -\dfrac{d[R]}{dt}$

Instantaneous rate of appearance of a product: $\quad \dfrac{d[P]}{dt}$

Half-life (the time taken for half of a reactant to be consumed or product to be formed): $\quad t_{1/2}$

The rate of formation of products P as a whole is equal to the negative of the rate of disappearance of the reactants R:

$$\frac{d[P]}{dt} = -\frac{d[R]}{dt}$$

But we must be careful in expressing the rate in terms of a single product or reactant. For example, consider the following reaction:

$$H_2(g) + I_2(g) \rightarrow 2HI(s).$$

Two moles of HI are produced at the rate of disappearance of 1 mole of H_2 per second. Thus, the rate of change of concentration, from **stoichiometry**, is

$$\frac{d[HI]}{dt} = -2\frac{d[H_2]}{dt}.$$

The rate at which H_2 disappears is the same as the rate at which I_2 disappears, and is half the rate at which HI appears. We cannot determine the order of the reaction from this net equation; we can only determine the relative magnitudes of the rates of products and reactants, no matter what the order for each participant might be.

7.9 Determination of Order

One of the most common ways of determining the rate law for a reaction is to measure the **initial rate of reaction** (e.g. formation of products) while varying the concentration of one of the reactants and holding the concentration of the other reactants constant.

$$\frac{d[P]}{dt} = k[A]^n[B]^m$$

Let us hold the concentration of B constant and express the rate as the rate of change of concentration of A:

$$-\frac{d[A]}{dt} = k'[A]^n$$

$$\log\frac{d[A]}{dt} = -n\log[A] - \log k'$$

By plotting the log of the initial reaction rate (d[A]/dt) versus log of the initial concentration [A], then the slope of the resulting straight line will be the order n.

Similarly, we may hold the concentration of A constant and vary the concentration of B to obtain a value for m.

$$-\frac{d[B]}{dt} = k''[B]^m$$

$$\log\frac{d[B]}{dt} = -m\log[B] - \log k''$$

It is important to know that the concentrations, and the rates, are determined at the initial conditions, that is, the rate of the reaction is measured as the slope of the tangent of the rate curve at $t = 0$.

The procedure can also be performed numerically, where we substitute values of reaction rate and concentration into the general rate law and solve simultaneous equations for the two unknowns n and m.

This method (either graphical or arithmetic) is called the **initial rate method**.

7.10 Determination of 1st Order Rate Law

An example for a simple **first-order reaction** is the decomposition of hydrogen peroxide to water and oxygen:

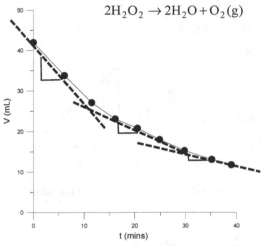

$$2H_2O_2 \rightarrow 2H_2O + O_2\,(g)$$

This is ordinarily a very slow reaction, but may be sped up by the addition of a catalyst. The concentration of H_2O_2 at any point is found by inhibiting the reaction by addition of dilute sulphuric acid, and the titration with $KMnO_4$. The volume of $KMnO_4$ used in the titration is an indication of the concentration of H_2O_2 left in the solution.

The slope of the tangent to the curve decreases as t increases, thus indicating that the rate of the reaction decreases with time. Since there is only one reactant, the decrease in reaction rate is therefore due to the decrease in concentration of H_2O_2. In general, the rate law would be written:

$$\frac{d[H_2O_2]}{dt} = k[H_2O_2]^n$$

For a first-order reaction, $n = 1$ and thus:

$$-\frac{d[H_2O_2]}{dt} = k[H_2O_2]$$

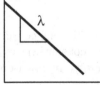

$$\frac{1}{[H_2O_2]}d[H_2O_2] = -kdt$$

$$\int_{[H_2O_2]_o}^{[H_2O_2]} \frac{1}{[H_2O_2]}d[H_2O_2] = -\int_0^t kdt$$

$$\ln[H_2O_2] - \ln[H_2O_2]_o = -kt$$

$$\ln[H_2O_2] = -kt + \ln[H_2O_2]_o$$

$$\log[H_2O_2] = -\frac{k}{2.303}t + \log[H_2O_2]_o \quad \text{(change of base to 10)}$$

$[H_2O_2]$ is proportional to the titration volume in this experiment, and so a plot of $\log V$ versus t is a straight line if the reaction is first order, and the value of k can be found from the slope of a plot of $\log V$ versus t.

94

7.11 Half-Life Method

The **half-life** $t_{1/2}$ of a reaction is the time for half of a reactant to be consumed, or half of a product to be formed. Consider previous case of a first-order reaction with one reactant:

$$2H_2O_2 \rightarrow 2H_2O + O_2(g)$$

Proceeding as before, we obtain

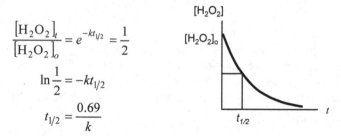

$$-\frac{d[H_2O_2]}{dt} = k[H_2O_2]$$

$$\ln[H_2O_2] = -kt + \ln[H_2O_2]_o$$

$$[H_2O_2]_t = [H_2O_2]_o\, e^{-kt}$$

The half-life $t_{1/2}$ is the time for one half of the reactant to be consumed:

$$\frac{[H_2O_2]_t}{[H_2O_2]_o} = e^{-kt_{1/2}} = \frac{1}{2}$$

$$\ln\frac{1}{2} = -kt_{1/2}$$

$$t_{1/2} = \frac{0.69}{k}$$

Either the rate constant k or the half-life (time measured for half the reactant to be consumed) can be determined from experimental data to give the other quantity. In a first-order reaction, the half-life is a constant and depends only on k.

7.12 Temperature

The rate of a chemical reaction depends on the frequency of collisions between atoms, ions or molecules, and also the fraction of which these particles have an energy greater than the activation energy. Increasing the concentration of reactants generally increases the frequency of collisions. Increasing the temperature also increases the collision frequency because the particles are moving faster and make more collisions per unit time. However, the most profound effect of an increase in temperature is to increase the fraction of particles with an energy above the activation energy.

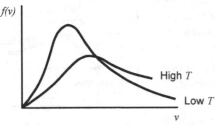

In gases, the distribution of velocities of molecules can be given by the Maxwell velocity distribution:

$$f(v) = 4\pi \left(\frac{m}{2\pi kT}\right)^{\frac{3}{2}} v^2 e^{-\frac{mv^2}{2kT}}.$$

At low temperature, a greater proportion of the molecules have a more narrow range of relatively low velocities compared to high temperature, where the range of velocities is more spread out, but shifted to higher values. Increasing the temperature therefore

• Shifts the molecular distribution of energies to a higher value
• Increases the proportion of molecules having higher energies

thus increasing the frequency and number of productive collisions per unit time and hence increasing the reaction rate.

The number of particles N with a total energy E was computed by Boltzmann using the **Maxwell velocity distribution** function:

$$N = Ce^{\frac{-E}{kT}}.$$

The rate constant for a specific reaction increases as the temperature increases, roughly in accordance with the **Arrhenius equation**:

$$k = Ae^{\frac{-E}{RT}},$$

where it can be seen that k is an exponential function of $1/T$. At room temperature, an increase of about 10°C results in an increase in reaction rate in the order of 2 to 3.

If E represents the activation energy, then this may be determined by experiment by plotting the logarithm of k versus $1/T$ over a range of temperatures.

7.13 Catalysts

The decomposition of hydrogen peroxide is a spontaneous, but slow reaction:

$$2H_2O_2 \rightarrow 2H_2O + O_2(g)$$

If powdered MnO_2 is added, the reaction proceeds much faster. Further, the MnO_2 molecules do not get used up during the reaction nor do they appear in the net equation. Substances which increase the rate of a reaction without appearing in the net reaction are called **catalysts** and the reaction is said to be **catalysed**.

The role of a catalyst is to alter the reaction mechanism by insertion of additional steps so that the activation energy is reduced, allowing the reaction rate to increase.

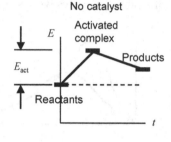

While the catalyst may be consumed in one reaction step, and regenerated in another, it does not appear in the net reaction.

Catalysts may be solids, liquids or gases. When the reaction being catalysed is all of the same phase (e.g. all liquid), it is called **homogenous catalysis**. When the reaction occurs at the surface of a catalyst, such as over the platinum bed of a **catalytic converter** in an automobile exhaust system, it is called **heterogeneous catalysis**.

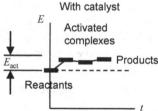

- A catalyst has no effect on the stoichiometry of the overall reaction.
- A catalyst does not affect the equilibrium position of the reaction. That is, the reaction proceeds to the same point of chemical equilibrium as it would without the catalyst. It just gets to equilibrium faster.
- A catalyst affects the reaction mechanism so as to lower the activation energy, thus increasing the number of molecules that can make a productive collision.

A negative catalyst decreases the rate of reaction and is called an **inhibitor**. The means by which this does so may not necessarily be to increase the activation energy, but is sometimes due to the inhibitor preventing another chemical present from acting as a catalyst. Auto catalysis, or **self-catalysis**, occurs when one of the products of a reaction acts as a catalyst for the reaction.

Enzymes act as catalysts in biological processes. Indeed, nearly all biological chemical reactions occur due to the action of catalysts.

8. Chemical Equilibrium

Summary

Law of chemical equilibrium:

$$K_c = \frac{[C]^c [D]^d}{[A]^a [B]^b}$$

Le Chatelier's principle:

When a chemical system is at equilibrium, and that equilibrium is disturbed in some way, then the system will change in a way so as to counteract the disturbance until equilibrium is re-established.

Gaseous equilibrium:

$$K_p = \frac{K_c}{RT} = \frac{P_c}{P_a P_b}$$

Solid/solubility equilibrium:

$$K_{sp} = \left[A^+(aq)\right]\left[B^-(aq)\right]$$

Common ion effect:

The presence of a common ion has reduced the solubility of a compound by several orders of magnitude.

8.1 Chemical Equilibrium

If a chemical reaction proceeds from reactants to products spontaneously, we may at first expect the process to continue until the reactants are all converted to product. This is not observed. Usually, reactions proceed until there is **chemical equilibrium**.

Reactants $\leftrightarrow$ Products

At equilibrium, the rate of forward reaction = rate of reverse reaction.

As the reaction proceeds, the reactants are consumed and the products are created. The concentration of the reactants decreases (unless they are continuously replenished) and the concentration of the products increases. The rate of a chemical reaction depends on the concentration of the reactants. Therefore if the concentration of reactants is steadily decreasing, then the rate of conversion of reactants to products (in moles/litre/second) is also decreasing. At the same time, the rate of the conversion of products back to reactants is increasing because the concentration of the products is increasing. At chemical equilibrium, the rates of both forward and reverse reactions are equal.

$$A + B \leftrightarrow C + D$$

At equilibrium, we cannot tell from this equation how many moles of A and B exist, and how many moles of the product C or D exist.

The concentrations of A and B start off at some particular values, then decrease and level off with time. The concentrations of C and D start from zero and increase and level off with time. At equilibrium, the concentrations of A, B, C and D are all constant.

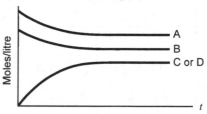

Although the balanced equation does not indicate the concentrations of reactants or products at any one time, they may be measured experimentally.

At equilibrium:

- The rate of formation of product C or D in moles per litre per second is equal to the rate of conversion of C and D back to molecules A and B.
- The concentrations of A, B and C, D are constant.

Chemical equilibrium is a **dynamic equilibrium**. Chemical activity still occurs at the equilibrium condition, but externally, we see no change in concentration of reactants or products.

It is important to know that **catalysts** do not alter the equilibrium state; they only serve to increase the rate of both forward and reverse reactions.

8.2 Law of Chemical Equilibrium

Consider the following single-step forward reaction:

$$A + B \rightarrow C + D$$

The rate law for this reaction is written:

$$\frac{d[C+D]}{dt} = k[A][B]$$

Now, consider the reverse reaction and its **rate law**:

$$C + D \rightarrow A + B$$

$$\frac{d[A+B]}{dt} = k'[C][D]$$

At equilibrium, the rate of forward reaction must equal the rate of the reverse reaction, and so:

$$k[A][B] = k'[C][D]$$

$$\frac{[C][D]}{[A][B]} = \frac{k}{k'} = K_c$$

That is, the ratio of the concentrations of the products over the reactants is a constant *at equilibrium*. A similar treatment can be written for more complex reaction steps.

More generally, it is found from experiment that for the general chemical equation $aA + bB \rightarrow cC + dD$ the **equilibrium constant** K_c is given by

$$K_c = \frac{[C]^c[D]^d}{[A]^a[B]^b} \quad \text{law of chemical equilibrium}$$

The magnitude of K_c indicates the extent of the reaction.

$K_c \leq 10^{-5}$	Forward reaction is negligible
$10^{-5} \leq K_c \leq 10^5$	Appreciable concentration of products and reactants
$K_c \geq 10^5$	Reverse reaction is negligible and almost complete conversion of reactants to products

The value of the equilibrium constant depends on the temperature and the chemical species involved. However, it does not depend on the starting concentrations of either reactants or products. Whatever concentrations of A, B, C and D there are, when equilibrium is reached, the ratio of the concentrations expressed in this way will have the same value at that temperature. The value of K also depends on how the balanced equation is written (i.e. the choice of coefficients). The equilibrium constant must be associated with a particular balanced equation.

Note that the equilibrium constant does not provide any information on how long it takes for equilibrium to be established. It only provides information on the extent to which equilibrium favours the products over the reactants.

8.3 Equilibrium Constant

For any chemical reaction, the ratio of concentrations raised to the power of the coefficients can be written as:

$$\frac{[C]^c [D]^d}{[A]^a [B]^b}$$

This is more formally called the **mass-action expression**. It is only at chemical equilibrium that this expression becomes equal to a constant value K_c (the subscript c indicates concentrations in moles/litre).

K_c depends on the temperature. The units of K_c depend on the powers to which the concentrations are raised.

$$N_2O_4 \rightarrow 2NO_2 (g)$$

$$K_c = \frac{[NO_2]^2}{[N_2O_4]} = \text{mol/L} \quad \longleftarrow$$

$$2CO_{(g)} + O_2 \rightarrow 2CO_2$$

$$K_c = \frac{[CO_2]^2}{[CO]^2 [O_2]} = \text{L/mol}$$

A particular value of K_c depends on the form of the balanced equation.

$$\frac{1}{2} N_2O_4 \rightarrow NO_2 (g)$$

$$K_c' = \frac{[NO_2]}{[N_2O_4]^{1/2}} = \text{mol}^{1/2} \text{ L}^{1/2}$$

$$= (K_c)^{1/2}$$

compare

Homogenous equilibrium occurs when all reactants and products are in the same physical state. **Heterogeneous equilibrium** occurs when reactants and products are in different physical states.

The concentrations of pure **solids** and **liquids** are constant. This is because if more *pure* solid or liquid is added, then the volume of that solid or liquid is also increased in the same proportion and so the number of moles per litre remains the same as before. The concentrations of these substances are removed from the equilibrium expression and are effectively incorporated into the value of K_c.

Gases cannot be removed from the equilibrium expression because their concentration changes during the reaction.

Writing a chemical equation in reverse inverts the value of K_c.

8.4 Le Chatelier's Principle

When a chemical system is at equilibrium, and that equilibrium is disturbed in some way, then the system will change in a way so as to counteract the disturbance until equilibrium is re-established. This is known as **Le Chatelier's principle**.

One way of disturbing the equilibrium condition is to add more of a reactant or a product. When more of a reactant is added, the forward reaction increases in rate (since reaction rate depends on the concentration) and more product is formed until the equilibrium condition, and the value of K_c, is restored. When more of product is added, then the concentration of the products increases, thus increasing the rate of the reverse reaction until the equilibrium condition is re-established. Similarly, if either a reactant or product is removed, then the forward or reverse reaction proceeds at a faster rate so as to restore the concentration of the removed species until the equilibrium condition is re-established.

Consider the effect of an increase in concentration of one reactant in the equilibrium expression:

$$N_2O_4 \rightarrow 2NO_2(g)$$

$$K_c = \frac{[NO_2]^2}{[N_2O_4]} = 0.87 \text{ at } 55°C$$

Increasing the amount of N_2O_4 makes the denominator of the expression larger, and makes the computed value of K_c smaller. The concentration of NO_2 must increase until the original value of K_c is re-established. Note that at the new equilibrium, the concentrations of both products and reactants will have increased but their ratio, as expressed by the equilibrium expression, remains the same. Looked at another way, introducing more reactant tends to increase the rate of the forward reaction until the equilibrium constant is re-established back to its former level.

Another way of disturbing the equilibrium condition is to alter the temperature. The value of K_c at equilibrium depends very much on the temperature. In an exothermic reaction, increasing the temperature tends to favour the reverse reaction. A new equilibrium is reached with a new value of K_c.

8.5 Summary of Le Chatelier's Principle

Summary of Le Chatelier's Principle:

Change	Effect on K_c	Shift in reaction
Increase in concentration of one reactant	Constant	Forward reaction
Decrease in concentration of one reactant	Constant	Reverse reaction
Increase in concentration of one of the products	Constant	Reverse reaction
Decrease in concentration of one of the products	Constant	Forward reaction
Pressure increase with volume decrease	Constant	Decrease in number of gaseous molecules
Pressure decrease with volume increase	Constant	Increase in number of gaseous molecules
Temperature increase (exothermic reaction)	Decrease	Reverse reaction
Temperature increase (endothermic reaction)	Increase	Forward reaction
Temperature decrease (exothermic)	Increase	Forward reaction
Temperature decrease (endothermic)	Decrease	Reverse reaction
Catalyst added	Constant	No shift, but rate of both forward and reverse reactions increased

8.6 Equilibrium in Gaseous Systems

For reactions involving gases, the law of chemical equilibrium can be written in terms of **partial pressures**. The amount of gas in a mixture can be measured in terms of its partial pressure. The total of all the partial pressures is the total pressure in the system. Consider the reaction

$$A(g) + B(g) \rightarrow C(g)$$

General gas equation: $PV = nRT$

Therefore, for A(g) $P_a V = nRT$ where P_a is the partial pressure of

$$P_a = \frac{n}{V} RT$$ gas A in volume V

$$= [A]RT$$ and similarly for B(g) and C(g).

Therefore, $K_c = \dfrac{[C]}{[A][B]}$

$$= RT \frac{P_c}{P_a P_b}$$

$$\frac{K_c}{RT} = \frac{P_c}{P_a P_b}$$ a constant.

$$= K_p$$

That is, a **gaseous equilibrium constant** K_p is the ratio of the partial pressures of the products over the partial pressures of the reactants.

Example: $N_2(g) + 3H_2(g) \rightarrow 2NH_3(g)$

$$K_p = \frac{P_{NH_3}^2}{\left(P_{N_2}\right)\left(P_{H_2}\right)^3}$$

$$K_c = \frac{[NH_3]^2}{[N_2][H_2]^3}$$

$$= \frac{\left(P_{NH_3}/RT\right)^2}{\left(P_{N_2}/RT\right)\left(H_2/RT\right)^3}$$

$$= \frac{P_{NH_3}^2}{\left(P_{N_2}\right)\left(P_{H_2}\right)^3}(RT)^2$$

$$= K_p (RT)^2$$

8.7 Solubility of Solids

Many chemical species dissolve in water. They can do so by a simple dissolution where the molecules just become separated from each other and are surrounded by water molecules, by dissociation into ions, and by chemical reaction with the water molecules.

Silver chloride, AgCl, has a low solubility in water. Those molecules that do dissolve, do so by forming ions. The **equilibrium constant** is therefore written as:

$$AgCl(s) \leftrightarrow Ag^+(aq) + Cl^-(aq)$$

Note that the chemical equation is written with the ions on the product side.

$$K_c = \frac{[Ag^+(aq)][Cl^-(aq)]}{[AgCl(s)]}$$

The concentration of a solid is a constant, and so is incorporated into the equilibrium constant to form a new constant, the **solubility product constant**, K_{sp}:

$$K_{sp} = [Ag^+(aq)][Cl^-(aq)] = 1.7 \times 10^{-10} \text{ at } 25°C$$

The **solubility product constant**, K_{sp}, applies to saturated aqueous solutions. The product of the ion concentrations is called the ion product. At equilibrium, the ion product equals K_{sp}. A low value of K_{sp} indicates that the concentration of ions in solution is low at equilibrium – in other words, the solubility of the compound is low.

The **solubility** of a compound in water is the number of moles of that compound that can be dissolved in 1 litre of water (at a specific temperature). The solubility of a compound can be obtained from the solubility product constant.

$$K_{sp} = [Ag^+(aq)][Cl^-(aq)]$$
$$= 1.7 \times 10^{-10} \text{ at } 25°C$$
$$s = [Ag^+(aq)] = [Cl^-(aq)]$$
$$1.7 \times 10^{-10} = s^2$$
$$s = 1.3 \times 10^{-5} \text{ mol/L}$$

From the balanced equation, 1 mole of AgCl will produce one mole of Ag^+ and 1 mole of Cl^- ions. Therefore in a 1 litre solution, there will be s moles of Ag^+ and s moles of Cl^- ions.

At equilibrium, the rate of **dissolution** is equal to the rate of **precipitation**.

In general, the solubility product constant is

$$M_nX_m \leftrightarrow nM^{n+}(aq) + mX^{n-}(aq)$$

$$K_{sp} = [M^{n+}(aq)]^n [X^{n-}(aq)]^m \quad \text{at solubility equilibrium}$$

Ionic solids (salts) and some molecular solids may dissociate into ions in aqueous solution.

8.8 Factors Affecting Equilibrium

A system tends towards a state of minimum **energy** and maximum **entropy**. The two tendencies can oppose or reinforce one another.

- If the forward reaction is exothermic $\Delta H < 0$ and the entropy change $\Delta S > 0$, then the forward reaction is favoured.
- If the forward reaction is endothermic $\Delta H > 0$ and the entropy change $\Delta S < 0$, then the reverse reaction is favoured.
- If the enthalpy change $\Delta H > 0$ and the entropy change $\Delta S > 0$, then the forward reaction is favoured at high temperature. When $\Delta H < 0$ and $\Delta S < 0$, the forward reaction is favoured at low temperature.

For a **solid** dissolving in a **liquid**, in saturated solution, the position of minimum energy can be on either side of the equation. The position of maximum entropy is on the dissolved side:

$$\text{Solid} \leftrightarrow \text{Saturated solution}$$

$$\text{entropy gain} \rightarrow$$

$$\leftarrow \text{energy minimum} \rightarrow$$

For the dissolving of a **gas** in a **liquid**, entropy decreases as the gas dissolves and so the tendency is to oppose the dissolving process, however, when a gas dissolves, the gas molecules, being attracted to the liquid molecules, enter a state of lower potential energy and so this favours the dissolving process.

$$\text{Gas} \leftrightarrow \text{Saturated solution}$$

$$\leftarrow \text{entropy gain}$$

$$\text{energy minimum} \rightarrow$$

It is important to realise that the **equilibrium constant** is not the same thing as the **solubility product constant**. The solubility product constant is an experimentally derived value that applies to the solubility of a compound in **aqueous solution**. The equilibrium constant applies to any chemical reaction, not just those in aqueous (water) solution.

Many chemical reactions occur in dilute aqueous solutions. In these cases, for reactions that consume, or produce, water molecules, the concentration of the water is considered constant and is incorporated into the equilibrium constant. This is because the amount of water consumed or produced is usually a small fraction of the overall total amount of water present.

8.9 Common Ion Effect

The solubility of an ionic solid in a solution containing a common ion is greatly reduced. This is because the concentration of the common ion appears in the expression for the solubility product constant.

$K_{sp} = 1.5 \times 10^{-9}$ for $BaSO_4$ in water . The solubility of this salt in water can be calculated and compared with the solubility in 0.1M Na_2SO_4.

Solubility in water:

$$K_{sp} = \left[Ba^{2+}(aq) \right]\left[SO_4^{2-}(aq) \right]$$
$$= (s)(s)$$
$$s^2 = 1.5 \times 10^{-9}$$
$$s = 3.9 \times 10^{-5} \, mol/L$$

Solubility in 0.1M Na_2SO_4:

$$K_{sp} = \left[Ba^{2+}(aq) \right]\left[SO_4^{2-}(aq) \right]$$
$$= (s)(s + 0.1)$$
$$= 1.5 \times 10^{-9}$$
$$(s + 0.1) \approx 0.1 \quad \text{since } s \text{ will be} \ll 0.1$$
$$1.5 \times 10^{-9} = s(0.1)$$
$$s = 1.5 \times 10^{-8} \, mol/L$$

The presence of the **common ion** has reduced the solubility of this compound by several orders of magnitude.

8.10 Precipitation

The **solubility product constant** K_{sp} can be used to predict whether or not a solid precipitate will form as a result of a chemical reaction between solutions.

The product of the concentrations of the ions in a solution is called the **ion product**. When the ion product is greater than the solubility product constant, **precipitation** will occur until the concentration of the ion reduces to the extent that the ion product becomes equal to K_{sp}.

Consider the mixing of 1 litre of 0.5M $Ca(NO_3)_2$ and 1 litre of 0.2M Na_2SO_4. We begin by writing the ionic equation of interest:

$$CaSO_4 \leftrightarrow Ca^{2+} + SO_4^{2-}$$

Note that the equation has been written with the ions on the product side of the equation so that the ion product can be related to K_{sp} directly, and not the inverse.

The final volume of the mixed solution is 2 litres, and therefore the concentrations of the original solutions are:

$$[Ca^{2+}] = 0.25$$
$$[SO_4^{2-}] = 0.1$$
$$IP = [Ca^{2+}(aq)][SO_4^{2-}(aq)] \quad \text{ion product}$$
$$= (0.25)(0.1)$$
$$= 0.025$$

K_{sp} for $CaSO_4 = 2.6\times10^{-4}$. Since the ion product is greater than K_{sp}, then $CaSO_4$ will form as a precipitate until the concentration of the Ca^{2+} and SO_4^{2-} ions fall until the ion product becomes equal to K_{sp}. That is, precipitation is a method of forcing out ions from the solution until the ion product becomes equal to K_{sp}, at which time the solid and the solution are in equilibrium. Precipitation has important economic consequences whereby the extraction of the desired product must be achieved as economically as possible. Precipitates are sometimes washed with a solution containing a **common ion** to further reduce their solubility.

8.11 Complex Ions

Some molecules or anions have a lone pair of electrons. This enables them to combine, or coordinate, with some transition metal ions to form a **complex ion**. For example, the molecule ammonia, NH_3, has a lone pair of electrons:

$$H : \overset{\cdot\cdot}{\underset{H}{N}} : H$$

NH_3 can coordinate to Ag^+, Co^{3+}, Zn^{2+}, Ni^{2+} to form complex ions:

$$Ag(NH_3)_2^+, Co(NH_3)_4^{2+}, Zn(NH_3)_4^{2+}, Ni(NH_3)_6^{2+}$$

The OH- ion also has a lone pair of electrons and can coordinate with Pb^{2+}, Zn^{2+}, Sn^{2+}, Al^{3+} :

$$Pb(OH)_4^{2-}, Zn(OH)_4^{2-}, Sn(OH)_4^{2-}, Al(OH)_4^-$$

The **ligand** (the species with the lone pair) arranges itself around the metal ion and the electron pairs fill up vacant energy levels to create a noble gas configuration for the metal ion. The electron orbitals that are associated with isolated atoms are rearranged to give **hybrid orbitals** to accommodate the electron pairs. The overall charge on the complex ion remains unchanged.

The equilibrium condition between the reactants and products for complex ion formation is called the **stability constant** K_{stab}.

$$Ni^{2+} + 4CN^- \rightarrow Ni(CN)_4^{2-} \qquad \text{Tetracyanonickelate}$$

$$K_{stab} = \frac{\left[Ni(CN)_4^{2-}\right]}{\left[Ni^{2+}\right]\left[CN^-\right]^4}$$

$$Pt^{2+} + 6NH_3 \rightarrow Pt(NH_3)_6^{2+} \qquad \text{Hexaamineplatinum (II)}$$

$$K_{stab} = \frac{\left[Pt(NH_3)_6^{2+}\right]}{\left[Pt^{2+}\right]\left[NH_3\right]^6}$$

$$Zn^{2+} + 4OH^- \rightarrow Zn(OH)_4^{2-} \qquad \text{Tetrahydroxozincate}$$

$$K_{stab} = \frac{\left[Zn(OH)_4^{2-}\right]}{\left[Zn^{2+}\right]\left[OH^-\right]^4}$$

9. Ionic Equilibrium

Summary

- Strong electrolytes dissociate completely into ions.
- Weak electrolytes only partially dissociate into ions.
- Non-electrolytes do not dissociate into ions when placed in solution.

Ion product of water:

$$K = \frac{\left[H^+\right]\left[OH^-\right]}{\left[H_2O\right]}$$

$$K_w = \left[H^+\right]\left[OH^-\right]$$

[H+] mol/L	pH	pOH
1	0	14
1×10^{-7}	7	7
1×10^{-14}	14	0

$$pH = -\log_{10}\left[H^+\right]$$

$$pOH = -\log_{10}\left[OH^-\right]$$

Acids – proton (H^+) donors

Bases – proton (H^+) acceptors

Acid equilibrium constant:

$$K_A = \frac{\left[H_3O^+\right]\left[A^-\right]}{\left[HA\right]}$$

Base equilibrium constant:

$$K_B = \frac{\left[BH^+\right]\left[OH^-\right]}{\left[B\right]}$$

Neutralisation: An acid on a base gives salt and water

Titration: The progressive addition of an acid to a base, or a base to an acid.

Hydrolysis: The chemical reaction of ions with water.

9.1 Electrolytes

When a substance dissolves, it may do so by simply separating from its neighbour molecules and become surrounded by water molecules, or the constituent molecules may break apart, or dissociate, and form negatively charged **anions** and positively charged **cations**. Such solutions conduct electricity by virtue of the movement of the ions under the influence of an applied electric field. Substances which dissolve to produce conducting solutions are called **electrolytes**. Substances which do not are called **non-electrolytes**.

- **Strong electrolytes** dissociate completely into ions.
- **Weak electrolytes** only partially dissociate into ions.
- **Non-electrolytes** do not dissociate into ions when placed in solution.

When an ionic substance dissociates into ions in an aqueous solution, the ions themselves are attracted to the polar water molecules and become **hydrated**. That is, a negative anion is surrounded by the positive (or H) end of water molecules and this hydrated ion can move around in solution under the influence of an electric field because as a whole, it is still negatively charged. Similarly, a positive cation is surrounded by the oxygen end of nearby water molecules and becomes hydrated, but the hydrated ion remains positively charged.

Both ionic and molecular compounds may be strong electrolytes. In the case of HCl, a molecular compound, we have

H$^+$ proton with positive charge showing one water molecule being attracted to it. Other water molecules may also be attracted to this proton.

$$HCl + H_2O \rightarrow H_3O^+ + Cl^-$$

The **hydronium ion**, H_3O^+, is a hydrated H^+ ion. For convenience, we omit the water of hydration and just write

$$HCl \rightarrow H^+ + Cl^-$$

Examples of ionic compounds which are strong electrolytes are NaCl, $MgCl_2$, KBr, $CuSO_4$, $LiNO_3$. Examples of molecular covalent compounds which are strong electrolytes are HCl, HBr, HNO_3. Examples of weak electrolytes are $HC_2H_3O_2$, HNO_2, NH_3, HCN.

The degree to which dissociation occurs depends on energy changes (in particular, the energy of hydration) and entropy considerations as well as the nature of the solvent, the concentration of the solute, and the temperature. The dissociation of HCl is an example of **ionic equilibrium**.

9.2 Ionisation of Water

Experiments indicate that pure **water** is a very poor conductor of electricity, but it does conduct, indicating the presence of ions. Water is a weak **electrolyte**. Water dissociates into ions according to

$$H_2O \leftrightarrow H^+ + OH^-$$

The H^+ ions attach themselves to other water molecules to form the **hydronium** ion H_3O^+, but here we will simply write the H^+ ions as if they exist on their own.

The degree of dissociation is very small, and so the concentration of the H^+ and OH^- ions is also very small.

The concentration of water, expressed as moles/litre, can be readily established from its molecular weight and its density:

$$m.w._{H_2O} = 16 + 2$$
$$= 18$$

One litre of water has a mass of 1000 g, therefore the number of moles n in a litre of water is:

$$n = \frac{1000}{18}$$
$$= 55.6 \text{ moles}$$

That is, water has a concentration of 55.6M.

Since the concentration of the H^+ and OH^- ions in a litre of water is exceedingly small, the concentration of H_2O is virtually a constant, and so the **ion product** K_w is formed:

$$K = \frac{[H^+][OH^-]}{[H_2O]}$$

$$K_w = [H^+][OH^-]$$

The concentration of H_2O (55.6M) is taken to be a constant and incorporated into K_w.

Experiments show that the **ion product of pure water** at 25°C is 1.0×10^{-14} mole²/litre². The concentrations of H^+ and OH^- ions are 1.0×10^{-7} moles/litre respectively.

A solution containing equal concentrations of H^+ and OH^- ions at these concentrations are said to be **neutral**.

$$[H^+] = 1 \times 10^{-7} \text{ mol/litre}$$
$$[OH^-] = 1 \times 10^{-7} \text{ mol/litre}$$

9.3 H+ and OH- Concentrations

Consider a 0.1M solution of HCl:

$$HCl \rightarrow H^+ + Cl^-$$

— because HCl dissociates completely to H+ and Cl- ions.

The concentration of H^+ ions will be 0.1M and Cl^- ions 0.1M. We can ignore the small amount of H^+ ions from the dissociation of the water solvent. But, the equilibrium between the H^+ and the OH^- ions in the solution must still be established such that

$$1 \times 10^{-14} = [H^+][OH^-]$$
$$= 0.1[OH^-]$$
$$[OH^-] = 1 \times 10^{-13} \text{ mol/litre}$$

That is, compared to neutral water, the concentration of the OH^- has been reduced from 1×10^{-7} to 1×10^{-13} due to a **common ion effect**.

Consider now an aqueous solution of 0.1M NaOH. Using the same calculation as above, we find that

$$NaOH \rightarrow Na^+ + OH^-$$
$$[H^+] = 1 \times 10^{-13} \text{ mol/litre}$$

The added H^+ or OH^- ions suppress or inhibit the dissociation of the water molecules (in accordance with Le Chatelier).

Because the changes in concentration of these ions can cover such a large range of magnitudes, it is convenient to express them on a logarithmic scale – the "p" scale. The **pH** is the concentration of H^+ ions found from:

$$pH = -\log_{10}[H^+]$$

The **pOH** is the concentration of OH^- ions found from
$$pOH = -\log_{10}[OH^-]$$

The pH is a measure of the potential to combine with H+ ions. A high pH means a high potential for H+ combination.

Since the logarithm is the index to which the base must be raised to equal the number, the pH is the index of the H^+ concentration. For example, for a neutral solution,

$$pH = -\log^{10}[H^+] = -\log_{10}(10^{-7})$$
$$= 7$$

It is easy to show that for a solution at equilibrium,

$$pH + pOH = 14$$

[H+] (mol/L)	pH	pOH
1	0	14
1×10^{-7}	7	7
1×10^{-14}	14	0

Say we have a 5M solution of HCl. What is the pH? $pH = -\log_{10} 5 = -0.7$

A negative pH is possible, but at high concentrations, the assumptions we make about the derivation of these equilibrium formulae breaks down.

9.4 Acids and Bases

A wide variety of substances can be generally classified into being either **acids**, **bases** or **salts**. Each of these has several distinctive characteristics when dissolved in water.

Acids	Bases
Electrical conductor	Electrical conductor
Sharp sensation on skin	Slippery feel
Sour taste	Bitter taste
Change the colour of litmus from blue to red	Change colour of litmus from red to blue.
React with some metals to form H_2 gas	
Dissolve carbonates to form CO_2 gas	

It was noticed that substances such as HCl, HNO_3, $HC_2H_3O_2$, H_2SO_4 and so on had the characteristics of acids. Substances such as $NaOH$, KOH, $Ca(OH)_2$, $Mg(OH)_2$ had the characteristics of bases. It was therefore postulated by **Arrhenius** (1859-1927) that acids are substances which dissolve in water to form H^+ ions, and bases are substances that dissolve in water to form OH^- ions:

$$\text{Acid} \qquad\qquad \text{Base}$$
$$HA \rightarrow H^+ + A^- \qquad BOH \rightarrow B^+ + OH^-$$

According to Arrhenius, all acids have the general formula HA and all bases have the general formula BOH.

A slightly more general definition of Arrhenius' scheme is that acids dissociate so as to increase the H^+ ions in water while bases dissociate and result in an increase in OH^- ions.

It was also known that when an acid was mixed with a base, the resulting solution lost its properties of either (except for being an electrical **conductor**) and formed a **salt** with the general form BA.

The degree of dissociation in water indicates the strength of the acid or the base:

$$HCl \rightarrow H^+ + Cl^- \qquad \text{Strong acid (equilibrium favours products)}$$

$$HC_2H_3O_2 \rightarrow H^+ + C_2H_3O_2^- \qquad \text{Weak acid (equilibrium favours reactants)}$$

$$NaOH \rightarrow Na^+ + OH^- \qquad \text{Strong base (equilibrium favours products)}$$

$$NH_3 + H_2O \leftrightarrow NH_4^+ + OH^- \qquad \text{Weak base (equilibrium favours reactants)}$$

$$HCl + NaOH \rightarrow NaCl + H_2O \qquad \text{acid + base = salt + water}$$

The Arrhenius definition of acids and bases remains the simplest and most popular despite it having some limitations in only dealing with H^+ and OH^- ions as the cause of acidic or basic properties of substances.

9.5 Bronsted–Lowry

Reactions can take place in other solvents besides water. H^+ and OH^- ions may not be involved at all, yet the reactions still have the characteristics of acids and bases. The **Bronsted–Lowry** definitions of acids and bases are

Acids – proton (H^+) donors
Bases – proton (H^+) acceptors

The Bronsted–Lowry classification is consistent with the Arrhenius definition of acids and bases but includes certain substances that act as bases which are not covered by the Arrhenius definition.

Consider the dissociation of HCl in water, and this time, we write the equation in full with the **hydronium ion**:

$$HCl + H_2O \rightarrow H_3O^+ + Cl^-$$

Now, let's focus our attention on the water molecule. The water molecule accepts a H^+ ion, a proton, to become a hydronium ion. That is, the water molecule, according to Bronsted and Lowry, acts as a base. At equilibrium, the reverse reaction also occurs:

$$HCl + H_2O \leftarrow H_3O^+ + Cl^-$$

Here, the hydronium ion donates a proton to the proceedings and the result is a water molecule. For all intents and purposes, the hydronium ion acts as an acid. In the Bronsted–Lowry scheme, the water molecule is the **conjugate base** of the hydronium ion. Or, the hydronium ion is the **conjugate acid** of the water molecule.

In the forward reaction, HCl donates a proton and becomes a Cl^- ion. HCl acts as an acid. In the reverse reaction, the Cl^- ion becomes HCl. That is, the Cl^- ion acts like a base, accepting a proton and becoming HCl. HCl is the **conjugate acid** of Cl^- and Cl^- is the **conjugate base** of HCl.

Every Bronsted–Lowry acid is accompanied by a Bronsted–Lowry base. The Bronsted–Lowry scheme has this extra dimension of creation of a **conjugate pair** compared to the Arrhenius theory:

$$HA + H_2O \rightarrow H_3O^+ + A^-$$
$$\text{acid} \quad \text{base} \quad \text{acid} \quad \text{base}$$

In the forward reaction, HA is the acid and H_2O is the base. For the reverse reaction, H_3O^+ is the acid and A^- is the base.

The degree of dissociation indicates the strength of the acid or the base. A strong Bronsted–Lowry acid implies a weak conjugate base. One of the greatest strengths of this system is that acid-base reactions that do not involve the OH^- ion can be accommodated.

9.6 Strength of Acids

The strength of an acid is described by its tendency to dissociate into ions and produce H^+. The **pH** scale does not necessarily indicate whether a solution contains a strong acid (or base); it only indicates the concentration of H^+ (and OH^-) ions which depends on both the intrinsic strength of the acid and the molar concentration used to make the solution.

HCl is considered an intrinsically strong acid because it almost completely dissociates into H^+ and Cl^- ions:

$$HCl \rightarrow H^+ + Cl^-$$

The **equilibrium constant** K_{HCl} for this reaction is very large ($\approx 10^7$), favouring he products. Acetic acid is a weak acid:

$$HC_2H_3O_2 \leftrightarrow H^+ + C_2H_3O_2^-$$

where we always know that H^+ is actually hydrated

The equilibrium constant for this reaction is very small (about 1.8×10^{-5}), favouring the reactants.

The general expression for the **acid equilibrium constant** is given as

$$HA + H_2O \leftrightarrow H_3O^+ + A^-$$

$$K = \frac{[H_3O^+][A^-]}{[HA][H_2O]}$$

$$K_A = \frac{[H_3O^+][A^-]}{[HA]}$$

since the concentration of H_2O is considered constant.

K_A provides a quantitative measure of the strength of an acid. That is, its tendency for dissociation to provide H^+ ions. pH is an indication of the concentration of H^+ ions in a particular solution. A highly concentrated solution of a weak acid may have a similar pH to a dilute solution of a strong acid.

Acid	K_A	Conjugate base
$HClO_4$	$\approx 10^{10}$	ClO_4^-
H_2SO_4	$\approx 10^9$	HSO_4^-
HCl	$\approx 10^7$	Cl^-
HSO_4	1.3×10^{-2}	SO_4^-
H_3PO_4	8×10^{-3}	$H_2PO_4^-$
$HC_2H_3O_2$	1.74×10^{-5}	$CH_3CO_2^-$
NH_4^+	6.3×10^{-10}	NH_3
NH_3	$\approx 10^{-30}$	NH_2^-

The strength of a base can be expressed in a similar manner by the **base equilibrium constant** K_B.

$$B + H_2O \leftrightarrow BH^+ + OH^-$$

$$K_B = \frac{[BH^+][OH^-]}{[B]}$$

These equilibrium constants can be expressed on a logarithmic scale:

$$pK_A = -\log_{10}[K_A]$$

$$pK_B = -\log_{10}[K_B]$$

For an acid and its conjugate base,

$$pK_A + pK_B = 14$$

9.7 Acid–Base Reactions

A few acids and bases are strong, but most are weak. Consider the dissociation of acetic acid $HC_2H_3O_2$, a weak acid with $K_A = 1.74 \times 10^{-5}$.

$$HC_2H_3O_2 + H_2O \leftrightarrow H_3O^+ + C_2H_3O_2^-$$

The concentration of H_3O^+ is found from:

$$K = \frac{[H_3O^+][C_2H_3O_2^-]}{[HC_2H_3O_2][H_2O]}$$

Strong acids:
HCl, HBr, HI, HNO_3,
H_2SO_4, $HClO_4$

$$[H_3O^+] = 1.74 \times 10^{-5} \frac{[HC_2H_3O_2]}{[C_2H_3O_2^-]}$$

If a solution is made with 0.1 moles of $HC_2H_3O_2$, then the concentration $[HC_2H_3O_2]$ is 0.1M. What then is the concentration of $[C_2H_3O_2^-]$? It must be small since the value of K_A implies a low incidence of dissociation. Now, let us add 0.1 moles of solid $NaC_2H_3O_2$. Since the salt, $NaC_2H_3O_2$ dissociates almost completely, the concentration of the acetate ion is now approximately 0.1M. Thus

$$[H_3O^+] = 1.74 \times 10^{-5} \frac{0.1}{0.1}$$

$$pH = 4.75$$

If to this solution we now add a small amount of a strong base, say 0.001 moles (a drop) of 2M NaOH, then the NaOH reacts with the acetic acid molecules: $HC_2H_3O_2 + NaOH \leftrightarrow H_2O + NaC_2H_3O_2$

The quantity $[HC_2H_3O_2]$ is reduced from 0.1M to 0.099M. The quantity $[C_2H_3O_2^-]$ increases to 0.11M (since the $NaC_2H_3O_2$ dissociates). The pH of this new solution is

$$pH = \log\left(1.74 \times 10^{-5} \frac{0.099}{0.11}\right) = 4.8$$

The $HC_2H_3O_2$ / $C_2H_3O_2^-$ solution is called a **buffer solution**, because of its ability to resist changes in pH. If a small amount of acid is added to the solution, this reacts with the acetate ion and the concentrations shift slightly the other way, and the pH remains almost unchanged.

If the concentration of the introduced acid or base becomes too large, then the **buffering capacity** of the solution is exhausted and the pH changes appreciably.

Note, without the common ion, addition of small amounts of acid or base can significantly alter the pH by several units.

9.8 Buffer Solutions

Buffer solutions have the important property of being able to maintain pH when disturbed by the addition of other acids or bases. Buffer solutions can be made in two ways:

1. Weak acid/salt of conjugate base (e.g. acetic acid and sodium acetate)

$$NaA_{(s)} \rightarrow Na^+ + A^-$$

$$HA + H_2O \leftrightarrow H_3O^+ + A^-$$

$$[H_3O^+] = K_A \frac{[HA]}{[A^-]}$$

$$pH = -\log\left(K_A \frac{[HA]}{[A^-]}\right)$$

There are two equilibrium reactions going on here. One is the almost complete dissociation of the salt into its ions, and the other is the partial dissociation of the acid. The [A⁻] is dominated by the molarity of the salt solution. The other reaction is the acid equilibrium. Because of the presence of the common ion A⁻, the equilibrium for this reaction lies on the reactants side of the equation and [HA] is therefore practically equal to the molarity of the acid.

Since [HA] is high (equilibrium to the left), any added OH⁻ from an introduced chemical will react with these ions:

$$OH^- + HA \rightarrow A^- + H_2O$$
$$\text{added} \quad \text{in buffer}$$

$$H_3O^+ + A^- \rightarrow HA + H_2O$$
$$\text{added} \quad \text{in buffer}$$

The added H_3O^+ or OH⁻ ions will be prevented from affecting the pH by combing with either the HA or A⁻ ions already in the solution.

2. Weak base/salt of conjugate acid (e.g. ammonia/ammonium chloride)

$$BH^+X^- \rightarrow BH^+ + X^-$$

$$B + H_2O \leftrightarrow BH + OH^-$$

$$[OH^-] = K_B \frac{[B]}{[BH^+]}$$

$$pOH = -\log\left(K_B \frac{[B]}{[BH^+]}\right)$$

$$pH = 14 - pOH$$

Complete dissociation of the salt into its ions results in the [BH⁺] being dominated by the molarity of the salt solution. Because of the presence of the common ion BH⁺, the equilibrium for the base reaction lies on the reactants side of the equation and [B] is therefore practically equal to the molarity of the base.

Any added H_3O^+ or OH⁻ ions will be prevented from affecting the pH by combing with either the B or BH⁺ ions already in the solution.

9.9 Indicators

Indicators are either a weak acid or a weak base with the property of exhibiting a change in colour depending on the pH of the solution. The molecular form of the acid HA has one colour, and the ionic form, A^-, another colour.

$$HA + H_2O \leftrightarrow H_3O^+ + A^-$$

Colour 1   Colour 2

If a small amount of indicator is added to a solution containing a large concentration of H_3O^+, then the equilibrium for the indicator solution shifts towards the reactants and Colour 1 dominates. If the indicator is added to a solution of large concentration of OH^-, then the equilibrium shifts towards the products and Colour 2 dominates.

Typically, a few drops of indicator are added to a solution whose pH is to be determined, and the resulting colour of the solution indicates the pH when read off against a colour chart.

For the indicator:

$$HA + H_2O \leftrightarrow H_3O^+ + A^-$$

$$K_A = \frac{\left[H_3O^+\right]\left[A^-\right]}{[HA]}$$

$$\left[H_3O^+\right] = K_A \frac{[HA]}{\left[A^-\right]}$$

Indicator	K_A	pH of colour change
Methyl orange	2×10^{-4}	3.7
Litmus	1×10^{-7}	7
Phenolphthalein	7×10^{-10}	9.1

An indicator changes colour when [HA] = [A^-],

$$\therefore \left[H_3O^+\right] = K_A$$

Indicators are useful when they exist in a solution whose pH is slowly changing. If the pH, or the [H_3O^+] of a solution changes so that [HA] becomes equal to [A^-], and the [H_3O^+] of the solution becomes equal to the K_A of the indicator and the colour changes.

It is important to know that an indicator does not necessarily change colour when [H_3O^+] = [OH^-]. Some indicators change colour on the basic side, while others change on the acidic side, and some at the neutral position, or **equivalence point**.

The **end point** occurs when the indicator changes colour. The reason why this is called the end point becomes clear when indicators are used in the process of **titration**.

9.10 Neutralisation

When an acid reacts with a base, the H^+ ions combine with the OH^- ions to form water, and the acidic and basic properties of the solution disappear along with the disappearing ions. This is **neutralisation**. That is, an acid on a base gives salt and water.

Consider 0.1M HCl added to 0.1M NaOH:

$$HCl + NaOH \rightarrow NaCl + H_2O$$

More specifically, in 0.1M solutions, we do not have HCl molecules and NaOH molecules. They are already dissociated into their constituent ions. The reaction is therefore actually

$$H_3O^+ + Cl^- + Na^+ + OH^- \rightarrow Na^+ + Cl^- + 2H_2O$$

The net reaction is therefore

$$H_3O^+ + OH^- \rightarrow 2H_2O$$

We know from experience that the products of this reaction are salt and water, or more specifically, Na^+ ions and Cl^- ions and water. That is, salty water, and not a mixture of acidic and basic solutions. But, how do we know this? Because in any aqueous solution at equilibrium, $K_w = 1 \times 10^{-14}$.

Equilibrium favours the products because the reverse reaction, the dissociation of water, has such a low **equilibrium constant** (or **ion product** in this case). The H_3O^+ ions combine with the OH^- ions to form water until such time as the product of the concentrations of H_3O^+ and OH^- ions fall to 1×10^{-14} moles/litre.

For strict neutralisation of a solution to occur, the initial concentrations of the H_3O^+ and OH^- ions must be similar – that is, the acid and the base must be of about the same strength. That is, if equal moles of acid and base are added, the final solution $[H_3O^+] = [OH^-] = 1 \times 10^{-7}$.

In a solution of acetic acid, very little dissociation occurs and so the reaction with a strong base in aqueous solution is:

$$HC_2H_3O_2 + Na^+ + OH^- \rightarrow H_2O + Na + C_2H_3O_2^-$$

$$HC_2H_3O_2 + OH^- \rightarrow H_2O + C_2H_3O_2^-$$

The resulting solution is not neutral, but slightly basic in the sense that one of the products, $C_2H_3O_2^-$, is a Bronsted–Lowry base. In general,

$$HA + OH^- \rightarrow H_2O + A^-$$

An interesting application of neutralisation is the use of antacids for relief of heartburn. In order to raise the pH in the stomach, sodium bicarbonate can be used to react with HCl in the gastric juices to give H_2O and $CO_2(g)$.

9.11 Titration

The progressive addition of an acid to a base, or a base to an acid, is called **titration**. Typically, when a base of known concentration is added to an acid of unknown pH, we say that the acid is titrated with the base. A known number of moles of base is added to the acid, either in solid form, or more commonly, as a solution via a burette.

As the base is added, the concentration of $[H_3O^+]$ changes due to neutralisation. If the pH is monitored during the process, a titration curve can be plotted showing pH against moles of added base.

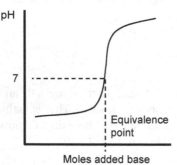

Moles added base

The moles of added base are measured usually by use of a burette. For the case of a strong acid and a strong base, there is a sharp change in pH at the equivalence point, where $[H_3O^+] = [OH^-]$. At or near this point, a very small addition of base results in a large increase in pH.

At the start of the titration, $[H_3O^+]$ is the molarity of the acid (for a strong acid). Before the equivalence point is reached, $[H_3O^+]$ is found from the initial moles of $[H_3O^+]$ less the moles of added OH^- (from the burette). At the equivalence point, $[H_3O^+] = [OH^-] = 1 \times 10^{-7}$. After the equivalence point, there is an excess of OH^-. Since we know how much OH^- is used to neutralise the acid, we know how much excess $[OH^-]$ exists after the equivalence point (from the burette reading) and the pH can be calculated from $14 - pOH$.

The change in pH of the solution as the base is added can be measured using a pH meter, or more conventionally, using an **indicator**. When the indicator changes colour, the **end point**, the titration is finished.

9.12 Choice of Indicator

During a titration, the indicator to be used should have an **end point** which corresponds to the **equivalence point**. At first sight, this might sound like that the only useful indicator would be one that has an end point at pH = 7. However, the choice of indicator depends on the nature of the reactants.

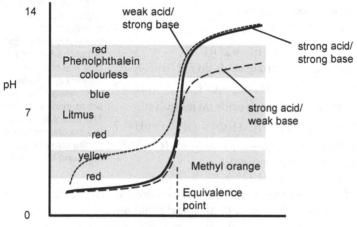

In the case of a strong acid being titrated with a strong acid, any of the three indicators above are suitable. However, for the case of a weak acid and a strong base, **Methyl orange** would not be suitable because it would change colour too early. **Phenolphthalein** would be a better choice because its end point (pH at which colour changes) more closely coincides with the equivalence point. For the case of a strong acid and a weak base, Methyl orange would be suitable, but not Phenolphthalein. **Litmus** is usually considered a poor choice because it changes colour over too wide a range of pH compared to other indicators. Weak acids are not usually used with weak bases in titration experiments.

A **universal indicator** is a mixture of other indicators designed so that different colours indicate different pH values.

In swimming pool maintenance, a few drops of phenol red (colour change from yellow to red over a pH range of 6.8 to 8.4) are used to measure pH. When the water is basic, the acid demand test involves putting a few drops of acid into the test mixture to determine how much acid is required to bring the solution back to the desired range of pH (usually 7.2 to 7.6).

9.13 Hydrolysis

Hydrolysis is the chemical reaction of ions with water. **Anions** may function as Bronsted-Lowry bases and receive a proton from the water molecule:

$$B^{n-} + H_2O \leftrightarrow OH^- + BH^{n-1}$$

Halide ions do not **hydrolyse** to any appreciable extent:

$$Cl^- + H_2O \leftarrow HCl + OH^- \qquad pH \approx 7$$

Oxide and sulphide ions hydrolyse:

$$O_2^{2-} + H_2O \rightarrow 2OH^- \qquad pH < 7$$

$$S^{2-} + H_2O \rightarrow HS^- + 2OH^-$$

Weak conjugate bases of strong acids do not hydrolyse:

$$ClO_4^- + H_2O \leftarrow HClO_4 + OH^- \qquad pH \approx 7$$

Strong conjugate bases of weak acids do hydrolyse:

$$C_2H_3O_2^- + H_2O \leftrightarrow HC_2H_3O_2 + OH^- \qquad pH < 7$$

Cations may function as Bronsted-Lowry acids and donate proton to the water molecule:

$$A^{n+} + H_2O \leftrightarrow H_3O^+ + A^{n-1}$$

Many elements occur in compounds as oxides and hydroxides. When hydrolysed, water molecules attach themselves to anions or cations formed by their dissociation in solution and these hydrolysed ions may act as acids or bases. The extent to which this occurs depends on the size of the atom and the charge.

Hydrated cations of groups I and II do not hydrolyse. Hydrated transition metal ions (Al_3^+, NH_4^+) hydrolyse to give acid solutions.

$$\left[Fe(H_2O)_6\right]^{3+} + H_2O \rightarrow H_3O^+ + \left[Fe(H_2O)_5OH\right]^{2+}$$

The **salts** of strong acids to not hydrolyse. For example, NaCl dissolved in water produces $\quad NaCl \rightarrow Na^+ + Cl^-$

$$Na^+ + H_2O \rightarrow \text{no reaction}$$
$$\qquad\qquad\qquad\qquad pH = 7$$
$$Cl^- + H_2O \rightarrow \text{no reaction}$$

The salts of weak acids and strong bases do hydrolyse. For example, $NaC_2H_3O_2$ (sodium acetate) when dissolved in water produces

$$NaC_2H_3O_2 \rightarrow Na^+ + C_2H_3O_2^-$$

$$Na^+ + H_2O \rightarrow \text{no reaction}$$

$$C_2H_3O_2^- + H_2O \leftrightarrow HC_2H_3O_2 + OH^- \qquad pH > 7$$

The salts of strong acids and weak bases do hydrolyse:

$$NH_4Cl \rightarrow NH^{4+} + Cl^-$$

$$Cl^- + H_2O \rightarrow \text{no reaction}$$

$$NH^{4+} + H_2O \rightarrow NH_3 + H_3O^+ \qquad pH < 7$$

The salts of weak acids and weak bases do hydrolyse but the resulting pH depends on the solubility constants K_A and K_B.

9.14 Simultaneous Equilibria

In the ideal case, only one equilibrium constant need be satisfied in a chemical reaction, but in practice, several equilibria are operating at the same time. For example, consider the hydrolysis reaction involving the dissociation of $(NH_4)_2S$ in solution:

$$NH_4^+ + H_2O \leftrightarrow NH_3 + H_3O^+$$
$$S^{2-} + H_2O \leftrightarrow HS^- + OH^-$$

There are two equilibria to be satisfied:

$$pK_{A(NH_4^+)} = 9.2$$
$$pK_{B(S^{2-})} = 0.2$$
$$pK_{B(S^{2-})} > pK_{B(NH_4^+)}$$
$$\therefore pH > 7$$

The quantitative analysis of hydrolysis involves the simultaneous equilibria of two reactions. For example consider the hydrolysis of the acetate ion:

$$C_2H_3O_2^- + H_2O \leftrightarrow HC_2H_3O_2 + OH^-$$

$$K_w = [H_3O^+][OH^-]$$
$$= 1 \times 10^{-14}$$

$$K_h = \frac{[OH^-][HC_2H_3O_2]}{[C_2H_3O_2^-]} \qquad \text{the \textbf{hydrolysis} \textbf{equilibrium constant}}$$

$$[OH^-] = \frac{K_w}{[H_3O^+]}$$

$$K_h = \frac{K_w[HC_2H_3O_2]}{[C_2H_3O_2^-]}$$

$$= \frac{K_w}{K_A} \qquad \text{The hydrolysis equilibrium constant is expressed in terms of the associated acid (or base) equilibrium constant.}$$

9.15 Complex Ions

Ligands are molecules or anions that have a lone pair of electrons. NH_3, H_2O and CN^- are examples. Some ligands combine with transition elemental metal ions – they become coordinated with the ion to form a **complex ion**.

$$[Cu(NH_3)_4]^{2+}$$

$$[Al(H_2O)_6]^{3+}$$

$$[Ag(CN)_2]^-$$

ligand
$$[M(L)_X]^{n+/-}$$
metal ion
coordination number

The **coordination number** usually equals twice the charge on the metal ion. The overall complex ion charge equals the sum of the cation and ligand charges. Complex ions may be anionic, cationic or neutral.

The properties of complex ions depend very much on the geometrical arrangement of molecules around the central atom. In general terms, the ligand provides electron pairs which find a place to attach themselves at available electron d orbitals around the central metal ion. However, this can happen in several ways. For example, consider the complex ion formed by Co and the ligands Cl^- and H_2O:

$$[Co(Cl)_4]^{2-}$$ coordination number 4

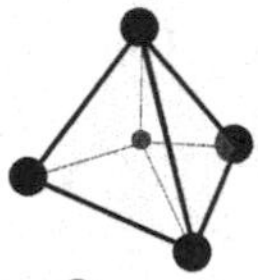

$$[Co(H_2O)_6]^{2+}$$ coordination number 6

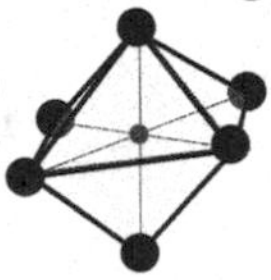

The lone pairs on the ligands face inwards towards and occupy d orbitals on the central metal ion. The reaction in water is

$$[Co(Cl)_4]^{2-} + 6H_2O \leftrightarrow [Co(H_2O)_6]^{2+} + 4Cl^-$$
blue   pink

Colours are often prominent in these compounds because the energy levels associated within the d orbitals are in the **visible light** range. Certain wavelength components of white light passing through solutions containing these ions are absorbed and the transmitted light that we see appears coloured.

10. Electronic Equilibrium

Summary

- **Oxidation** occurs at the **anode**. Neutral atoms being oxidised **lose** their electrons and become positively charged **cations** which tend to move towards the **cathode**.

$$4Fe(s) \rightarrow 2Fe_2^{2+}(aq) + 4e^-$$ **Oxidation** occurs when the oxidation

$$Cu(s) \rightarrow Cu^{2+}(aq) + 2e^-$$ number of an atom is increased (loss

$$Zn(s) \rightarrow Zn^{2+}(aq) + 2e^-$$ of electrons).

$$H_2^{2+}(g) \rightarrow 2H^+ + 2e^-$$

- **Reduction** occurs at the **cathode**. Positively charged ions in solution take up the electrons coming from the oxidation reaction and become neutral atoms. The remaining negatively charged **anions** tend to move towards the **anode**.

$$3O_2 + 4e^- \rightarrow 2O_3^{2-}$$

$$2Ag^{2+} + 2e^- \rightarrow 2Ag(s)$$ **Reduction** occurs when the oxidation

$$2H^+ + 2e^- \rightarrow 2H_2(g)$$ number of an atom is decreased (gain

$$Cu^{2+} + 2e^- \rightarrow Cu(s)$$ of electrons).

10.1 Oxidation and Reduction

Oxidation and reduction reactions involve the **equilibrium** of a competition for electrons between atoms and ions. It is **electronic equilibrium**. To see how this comes about, we need to look at some numbers that quantify what drives this type of reaction.

The **oxidation number**, or **oxidation state**, is a quantity that describes the number of electrons that seem to be gained or lost by an atom when a chemical bond is formed. During a chemical reaction, **oxidation** occurs when the oxidation number of an atom is increased (loss of electrons). **Reduction** occurs when the oxidation number of an atom is decreased (gain of electrons). That is, the term reduction indicates a reduction in oxidation number.

Consider a simplified version of the oxidation of iron:

$$4Fe(s) + 3O_2(g) \rightarrow 2Fe_2O_3(s)$$

The **oxidation half-reaction** is

$$4Fe(s) \rightarrow 2Fe_2^{2+} + 4e^-$$

In this half-reaction, the oxidation number of the Fe atoms increases from 0 to 2+ (loses electrons)

The **reduction half-reaction** is

$$3O_2 + 4e^- \rightarrow 2O_3^{2-}$$

In this half-reaction, the oxidation number of the O atoms decreases from 0 to 2− (gains electrons).

The terms oxidation and reduction are historical in origin. They arose from the common observation that some reactants combine with oxygen (and so are said to be **oxidised**) while others lose oxygen atoms (and so are said to be **reduced**). Now it is recognised that it is the loss and gain of electrons that is the important feature of such reactions.

The **oxidising agent**, or **oxidant**, is the substance that is reduced. In the example above, oxygen is the oxidising agent since oxygen oxidises the iron. The **reducing agent**, or the **reductant**, is the substance that is oxidised. A **redox couple** is an oxidant and its complementary reductant. In the above example, the redox couples are written

$$Fe/Fe_2^{2+} \quad \text{and} \quad O_2/O_3^{2-}$$

Oxidant/Reductant
(or Reduction/Oxidation)

The processes of oxidation and reduction do not occur in all chemical reactions. In many reactions, the oxidation numbers of the participating atoms do not change. Examples of **non-redox reactions** are some precipitation reactions, acids reacting with bases, acids and oxides, hydroxides, sulphides and carbonates. In **acid-base reactions**, it is more an exchange of protons that is important, not electrons.

10.2 Redox Reaction

Consider the redox reaction: $Cu(s) + 2Ag^{2+}(aq) \rightarrow Cu^{2+}(aq) + 2Ag(s)$

We can separate out the oxidation and reduction processes physically using separate beakers of solution, a wire for electrons to flow along, and a salt bridge for ions to move across and electrodes where reactions take place.

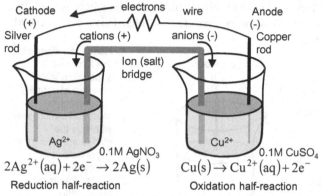

The salt bridge is a tube containing a salt solution (e.g. $NaNO_3$) and allows negatively charged anions to move into the anode solution and positively charged cations to move into the cathode solution while at the same time preventing bulk mixing of the solutions.

$2Ag^{2+}(aq) + 2e^- \rightarrow 2Ag(s)$ $Cu(s) \rightarrow Cu^{2+}(aq) + 2e^-$

Reduction half-reaction   Oxidation half-reaction

The electrode where reduction occurs is called the **cathode**. Silver has a greater affinity for **electrons** than copper. So, the Ag^{2+} ions in the solution combine with the free electrons in the cathode to form solid silver. This loss of free electrons at the cathode represents a net positive charge. The Ag^{2+} ions are pulling the electrons off the silver rod and forming solid silver at the cathode. But the consumption of Ag^{2+} ions in the solution results in a build-up of negative charge (the NO_3^- ions), and if this is not balanced in some way, the reaction will come to a stop as the ions in the solution repel any more electrons arriving at the cathode wanting to take part in the reaction.

The electrode where oxidation occurs is called the **anode.** The copper rod at the anode dissolves away to form Cu^{2+} ions and free **electrons**. These free electrons travel along the wire and are consumed at the cathode. The build-up of positively charge Cu^{2+} ions would eventually overcome the affinity for electrons by the silver and the reaction would come to a stop.

If the NO_3^- **anions** are allowed to flow from the cathode solution to the anode solution, and the Cu^{2+} **cations** flow from the anode solution to the cathode solution, then build-up of excess charge would be prevented and the reactions at each electrode would proceed spontaneously.

The redox reactions, arranged in this way, constitute an **electrochemical cell** (or **galvanic cell**) since a **voltage** is produced between the cathode and the anode. The sign of the anode and cathode is determined by the flow of electrons in the external wire, not the ion flow in the solutions.

10.3 Single-Cell Redox Reaction

Consider the redox reaction

$$Cu(s) + 2Ag^{2+}(aq) \rightarrow Cu^{2+}(aq) + 2Ag(s)$$

We have seen that this is an oxidation-reduction reaction whereby each half-reaction can be arranged to occur in separate solutions. This reaction can also occur in a single solution.

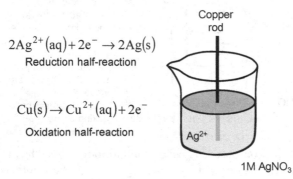

$$2Ag^{2+}(aq) + 2e^- \rightarrow 2Ag(s)$$
Reduction half-reaction

$$Cu(s) \rightarrow Cu^{2+}(aq) + 2e^-$$
Oxidation half-reaction

Copper rod

Ag²⁺

1M AgNO₃

Redox reactions of this kind find application in the silvering of copper conductors in electrical switch gear whereby the oxidation (dulling) of copper due to atmospheric contaminants is to be avoided.

When a copper rod is dipped into a silver nitrate solution, silver metal is spontaneously deposited on the rod. The solution turns blue. Electron transfer occurs from the copper to the silver ions in the vicinity of the rod.

Another good example of a single-cell reaction is when a zinc rod is dipped into hydrochloric acid.

Zinc rod

$$2H^+ + 2e^- \rightarrow 2H_2(g)$$
Reduction half-reaction

$$Zn(s) \rightarrow Zn^{2+}(aq) + 2e^-$$
Oxidation half-reaction

H⁺   H₂(g)

1M HCl

The metal dissolves in the acid, producing H_2 gas. The same thing happens with Mg, Al, Fe and Ni, but not with Cu, Ag and Au. The reason for some metals being able to be dissolved in HCl and others not can be found by a consideration of **standard electrode potentials**.

10.4 Standard Hydrogen Electrode

Consider the redox reaction

$$Cu^{2+}(aq) + H_2(g) \rightarrow Cu(s) + 2H^+(aq)$$

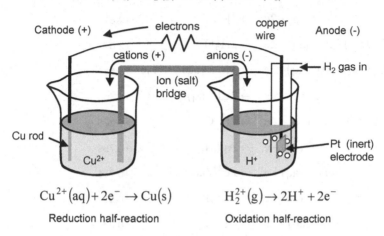

Cathode (+) ← electrons copper wire Anode (-)

cations (+) anions (-)

Ion (salt) bridge

← H_2 gas in

Cu rod

Cu^{2+}

H^+

Pt (inert) electrode

$$Cu^{2+}(aq) + 2e^- \rightarrow Cu(s)$$ $$H_2^{2+}(g) \rightarrow 2H^+ + 2e^-$$

Reduction half-reaction   Oxidation half-reaction

Cu^{2+}/Cu Redox couple $H^+/H_2(g)$ Redox couple

When the H_2 gas is at 1 atm and the H^+ solution is 1 M, the H^+/H_2 redox couple is referred to as a **standard hydrogen electrode** and is assigned a 0V potential. The potential difference between the hydrogen standard cell and the other electrode is called the **standard electrode potential** E° for the half-reaction occurring at that other electrode.

Standard electrode potentials are provided from the point of view of **oxidation** at the hydrogen standard electrode. That is, positive values of standard electrode potentials for a redox reaction indicate that the reaction will be one of **reduction** when the other half-cell is the standard cell. When the standard electrode potential for a redox reaction is listed as negative, it indicates that this reaction will be an **oxidation** reaction if the other electrode is the standard hydrogen electrode.

The standard electrode potential E° for the Cu^{2+}/Cu half-reaction above is +0.34V and so the copper is reduced and the hydrogen is oxidised. The value of E° for Zn is −0.76V and so if a zinc electrode and solution were used in place of the copper, the zinc would be oxidised and the hydrogen would be reduced.

10.5 Standard Electrode Potentials

Standard electrode potentials $E°$ for some **reduction half-reactions** where gases are at 1 atm and solutions are 1M:

$$F_2(g) + 2e^- \rightarrow 2F^- \qquad +2.87$$

$$O_3(g) + 2H^+(aq) + 2e^- \rightarrow 2H_2O \qquad +2.07$$

$$MnO_4^{2+}(aq) + 8H^+(aq) + 5e^- \rightarrow Mn^{2-}(aq) + 4H_2O \quad +1.51$$

$$Cl_2(g) + 2e^- \rightarrow 2Cl^-(aq) \qquad +1.36$$

$$O_2(g) + 4H^+(aq) + 4e^- \rightarrow 2H_2O \qquad +1.229$$

$$Ag^+(aq) + e^- \rightarrow Ag(s) \qquad +0.799$$

$$Fe^{3+}(aq) + e^- \rightarrow Fe^{2+}(aq) \qquad +0.771$$

$$Cu^+(aq) + e^- \rightarrow Cu(s) \qquad +0.52$$

$$Cu^{2+}(aq) + 2e^- \rightarrow Cu(s) \qquad +0.337$$

$$Sn^{4+}(aq) + 2e^- \rightarrow Sn^{2+}(aq) \qquad +0.15$$

$$S(aq) + 2H^+ + 2e^- \rightarrow H_2S(g) \qquad +0.141$$

$$2H^+(aq) + 2e^- \rightarrow H_2^{2+}(g) \qquad 0$$

$$Pb^{2+}(aq) + 2e^- \rightarrow Pb(s) \qquad -0.126$$

$$Sn^{2+}(aq) + 2e^- \rightarrow Sn(s) \qquad -0.136$$

$$Fe^{2+}(aq) + 2e^- \rightarrow Fe(s) \qquad -0.440$$

$$Zn^{2+}(aq) + 2e^- \rightarrow Zn(s) \qquad -0.763$$

$$Al^{3+}(aq) + 3e^- \rightarrow Al(s) \qquad -1.66$$

$$Mg^{2+}(aq) + 2e^- \rightarrow Mg(s) \qquad -2.37$$

$$Na^+(aq) + e^- \rightarrow Na(s) \qquad -2.714$$

$$Ca^{2+}(aq) + 2e^- \rightarrow Ca(s) \qquad -2.89$$

$$K^+(aq) + e^- \rightarrow K(s) \qquad -2.925$$

$$Li^+(aq) + e^- \rightarrow Li(s) \qquad -3.045$$

10.6 Spontaneous Redox Reactions

A very useful application of standard electrode potentials is the determination of whether a redox reaction will proceed spontaneously or not. Consider the copper and zinc redox couples:

$$Cu^{2+}/Cu \qquad\qquad Zn^{2+}/Zn$$

$$E^\circ = +0.34\,V \qquad E^\circ = -0.76\,V$$

There are four possible reactants and two possible reactions:

1 $Cu^{2+}_{(ox)} + Zn^{2+}_{(ox)}$ No reaction (simultaneous oxidation)

2 $Cu^{2+}_{(ox)} + Zn_{(red)}$ $Cu^{2+}(aq) + Zn(s) \rightarrow Cu(s) + Zn^{2+}(aq)$

3 $Cu_{(red)} + Zn^{2+}_{(ox)}$ $Cu_{(red)} + Zn^{2+}_{(ox)} \rightarrow Cu^{2+}(aq) + Zn(s)$

4 $Cu_{(red)} + Zn_{(red)}$ No reaction (simultaneous reduction)

The first and fourth possibilities cannot happen since we must have complementary oxidation and reduction processes happening. To determine which of the second and third possibilities will happen spontaneously when these chemicals are in an **electrochemical cell**, we add the **standard electrode potentials**. However, the standard electrode potentials are expressed in terms of reduction reactions, and so we must reverse the sign for a corresponding oxidation reaction.

For the second possibility, the electrode potential for the oxidation of copper in solution would be –0.34V. This is added to –0.76 for the reduction of zinc to obtain –1.1V. Since this total is a negative voltage for the cell as a whole, we say that this reaction will not proceed of its own accord.

For the third possibility, the electrode potential for the reduction of copper in solution is +0.34V. This is added to +0.76 for the oxidation of zinc to obtain +1.1V. Since this total is a positive value, we say that this reaction will proceed of its own accord and what's more, we would measure a **potential difference** of +1.1V across the terminals of the cell.

Standard electrode potentials for a redox reaction indicate the **reduction** potential of the half-reaction. Generally speaking, a redox reaction will proceed spontaneously if the **oxidant** (the one that is reduced) from the redox couple with the higher value of E° reacts with the **reductant** of the other.

Redox couple	$E^\circ V$
Oxidant / Reductant	Higher
Oxidant / Reductant	Lower

Reaction occurs

10.7 Oxidation Numbers

Although we can predict which redox reactions will occur spontaneously from a consideration of **standard electrode potentials**, we have yet to consider why the standard potentials are what they are. Consider the redox reaction

$$Cu(s) + 2Ag^{2+}(aq) \rightarrow Cu^{2+}(aq) + 2Ag(s)$$ Redox reaction

$$2Ag^{2+}(aq) + 2e^- \rightarrow 2Ag(s)$$ Reduction half-reaction

$$Cu(s) \rightarrow Cu^{2+}(aq) + 2e^-$$ Oxidation half-reaction

From the above, we can conclude that silver ions have a stronger affinity for attracting **electrons** than do copper ions, and so the silver ions are reduced to a solid by taking up electrons and the solid copper is oxidised by releasing electrons, rather than the reverse. It is the relative affinity of ions to attract electrons that places a particular element in the list of standard potentials – which as will be remembered is presented from the point of view of reduction half-reactions.

This affinity for electrons by atoms is something we have met before: **electronegativity**. It was found that a practical measure of electronegativity is the method of electron counting by **oxidation numbers**. In a chemical reaction, when the oxidation number of an atom (i.e. its apparent charge) has increased, the atom is said to have been **oxidised**. When the oxidation number of an atom has decreased, the atom is said to have been **reduced**.

In the above reactions, the oxidation number of copper is initially 0. After reaction, the copper atoms have apparently lost two electrons and so its oxidation number is now 2+ (its apparent charge), and is thus an increase. The oxidation number of silver in the Ag^{2+} ions is initially 2+, and after reaction is 0. The oxidation number has decreased and the silver is therefore reduced. All this comes down to a competition for the attraction of electrons.

- **Oxidation** represents a loss of electrons or increase in oxidation number.

- **Reduction** represents a gain of electrons or decrease in oxidation number.

Oxidation numbers allow us to recognise oxidation-reduction reactions and are also useful for balancing the equations for these reactions. Changes in oxidation number do not necessarily indicate that an atom has gained or lost an electron, or has gained or lost net electric charge, even though we use them in this way for the purpose of establishing the numbers. Rather, they indicate how many electrons are involved for the atom in a chemical reaction.

10.8 Balancing Redox Half-Reactions

In redox reactions, both atoms and charges must be conserved. A convenient method of balancing redox equations is to use the equations for the half-reactions.

Consider the reaction between copper and nitric acid, HNO_3. The redox couples involved are

$Cu^{2+}/Cu(s)$ $H^+, NO_3^-/NO(g)$

$E^° = +0.34V$ $E^° = +0.96V$

Since $E^°$ for the nitric acid is more positive, then the copper must undergo oxidation:

and the NO_3 must undergo reduction:

$Cu(s) \rightarrow Cu^{2+}(aq) + 2e^-$ $4H^+ + NO_3^- + 3e^- \rightarrow NO(g) + H_2O$

Both these half-reactions are balanced, but they now need to be combined to form the net reaction, which is required also to be balanced. The procedure is to balance the charge. The lowest common factor in this example is 6, and so

$3Cu(s) \rightarrow 3Cu^{2+}(aq) + 6e^-$ $8H^+ + 2NO_3^- + 6e^- \rightarrow 2NO(g) + 4H_2O$

These half-reactions can now be combined:

$$3Cu + 8H^+ + 2NO_3^- \rightarrow 3Cu^{2+} + 2NO(g) + 4H_2O$$

Not all reactions can be balanced so easily. A general method is to

1. Balance for atoms other than O and H
2. Balance for O by adding water
3. Balance for H by adding H^+
4. Add OH^- ions to both sides for basic solution to form water with the H^+ added from (3)
5. Balance for charge by adding electrons

It should be noted that balancing a half-reaction by adjusting coefficients does not alter the electrode potential for that half-reaction.

10.9 Balancing Redox Reactions with Oxidation Numbers

Another method of balancing redox reactions is by the use of oxidation numbers. Consider the reaction between copper and nitric acid HNO_3. The redox couples involved are:

$Cu^{2+}/Cu(s)$ $\qquad\qquad$ $NO_3^-/NO(g)$

$E^o = +0.34$ V $\qquad\qquad$ $E^o = +0.96$ V

Since E^o for the nitric acid is more positive, then the copper must undergo oxidation. The unbalanced equation is:

$$Cu + NO_3^- \rightarrow Cu^{2+} + NO(g)$$

Next, we write the oxidation numbers to each element in the reaction:

$$Cu + NO_3^- \rightarrow Cu^{2+} + NO(g)$$
$$0 \quad 5+,2- \quad 2+ \quad 2+,2-$$

Cu has undergone a change in oxidation number from 0 to 2+. An increase in oxidation number indicates oxidation. The change in oxidation number in this element is +2.

Rules for oxidation numbers

a) In free, uncombined, elements, the oxidation number of each atom is set to be 0.

b) In compounds involving hydrogen, the oxidation number of hydrogen is usually 1+.

c) In compounds involving oxygen, the oxidation number of O is usually 2−.

d) The sum of all the oxidation numbers of all atoms in an ion is equal (in both magnitude and sign) to the charge on the ion.

e) The sum of all the oxidation numbers of all atoms in a neutral molecule is 0.

N has undergone a change in oxidation number of 5+ to 2+. The change in oxidation number is −3. A decrease in oxidation number indicates reduction. There is no net change in oxidation number for the oxygen atoms.

We now introduce a further rule for oxidation numbers, and that is, the net *change* in oxidation number for the overall reaction is to be zero. That is, the overall *change* in oxidation number for the left side of the reaction has to be equal to the negative of the overall *change* oxidation number for the right side. The only way to do this here is to multiply the Cu atoms by 3 and the NO_3^- ions by 2. That is, operating on the changes in oxidation numbers, we have: $3(+2) + 2(-3) = 0$ Thus,

$$3Cu + 2NO_3^- \rightarrow 3Cu^{2+} + 2NO(g).$$

We need to add H_2O to get a balance for oxygen atoms:

$$3Cu + 2NO_3^- \rightarrow 3Cu^{2+} + 2NO(g) + 4H_2O$$

And finally, add H^+ ions to balance for H atoms:

$$3Cu + 8H^+ + 2NO_3^- \rightarrow 3Cu^{2+} + 2NO(g) + 4H_2O$$

10.10 Electrochemical Cell

An **electrochemical cell, or galvanic cell**, is a redox reaction that proceeds spontaneously and has an **electrical potential difference** at its electrodes. This potential difference, when measured at a negligible current draw, is called the **emf** of the cell E_{cell}.

$$E_{cell} = E_{cathode} - E_{anode}$$

reduction   oxidation

When the half-cells are standard electrodes, the standard emf of the cell is given by

$$E^o_{cell} = E^o_{cathode} - E^o_{anode}$$

Example of electrochemical cell:

$$Cu^{2+}/Cu \qquad Fe^{3+}(aq)/Fe^{2+}(aq), Pt$$

$$E^o = +0.34V \qquad E^o = +0.77V$$

All at 1M concentrations

Since E^o for the iron reaction is higher than that for copper, then the Fe^{3+} is undergoing reduction (it is the oxidant) and so is the **cathode** for the cell. Copper is being oxidised and so is the **anode**.

$$E^o_{cell} = E^o_{cathode} - E^o_{anode}$$
$$= 0.77 - 0.34$$
$$= +0.43V$$

Note: we do not have to "balance" the half-equations to determine the overall cell potential. For example, the **standard electrode potential** of the copper half cell is still +0.34 if we write

$$2Cu^{2+} + 4e^- \rightarrow 2Cu(s)$$

The potentials do depend on the concentration of the species, but not the number of atoms that participate in the overall reaction.

In an electrochemical cell that does not have a connection between the anode and cathode, the redox reaction proceeds until the build-up of positive charge on the cathode prevents any more cations from moving from the anode and vice versa and the reaction stops. The system has acquired **electrical potential energy** as a result of the reaction proceeds. The reduction of energy of the system as a whole (i.e. the chemical reaction tending to send the system towards a lower energy state) appears as mechanical work or heat by virtue of the current flow.

10.11 EMF vs. Concentration

The potential of an electrochemical cell depends upon the concentration of the reactants. Standard potentials are given for 1M solutions. As the cell operates, the concentrations of the reactants change. The dependence of the cell potential on concentration is given by the **Nernst equation**.

$$pOx + ne^- \rightarrow qRed$$

$$E = E^\circ + \frac{RT}{nF} \ln \frac{[Ox]^p}{[Red]^q}$$

R - universal gas constant
T - temperature (K)
F - Faraday's constant 96487 C/mol

$$= E^\circ + \frac{0.059}{n} \log_{10} \frac{[Ox]^p}{[Red]^q} \quad \text{at } 25^\circ C$$

Consider the electrochemical cell at 25°C:

$$Pt \mid Sn^{4+}(0.5M), Sn^{2+}(1.5M) \parallel Fe^{3+}(1.0M), Fe^{2+}(0.5M) \mid Pt$$

Anode   Cathode

$$Sn^{4+} + 2e^- \leftrightarrow Sn^{2+} \qquad\qquad Fe^{3+} + e^- \leftrightarrow Fe^{2+}$$

$$E_{Sn} = E^\circ_{Sn} + \frac{0.059}{2} \log_{10} \frac{[Sn^{4+}]}{[Sn^{2+}]} \qquad E_{Fe} = E^\circ_{Fe} + \frac{0.059}{1} \log_{10} \frac{[Fe^{4+}]}{[Fe^{2+}]}$$

$$= 0.15 + \frac{0.059}{2} \log_{10} \frac{[0.5]}{[1.5]} \qquad\qquad = 0.77 + \frac{0.059}{1} \log_{10} \frac{[1.0]}{[0.5]}$$

$$= 9.14V \qquad\qquad\qquad\qquad = 0.79 \text{ V}$$

Since $E_{Fe} > E_{Sn}$, $Fe^{3+}|Fe^{2+}$ is the cathode and the cell reaction is

$$2Fe^{3+} + Sn^{2+} \rightarrow 2Fe^{2+} + Sn^{4+}$$

$$E_{cell} = 0.79 + -0.14$$

$$= 0.65V$$

When the electrode is a metal in contact with its cation (e.g. $Cu^{2+}|Cu$), then

$$Cu^{2+} + 2e^- \leftrightarrow Cu_{(s)}$$

$$E = E^\circ + \frac{0.059}{2} \log_{10} [Cu^{2+}]$$

When the electrode is a gas in contact with its ion (e.g. $H^+|H_2(g)$), then

$$2H^+ + 2e^- \leftrightarrow H_2(g)$$

$$E = E^\circ + \frac{0.059}{2} \log_{10} \frac{[H^+]^2}{(P_{H2})} \longrightarrow \text{Pressure in atm}$$

10.12 Electronic Equilibrium

Thus far we have considered the conditions under which an oxidation-reduction reaction will proceed. Consider an electrochemical cell:

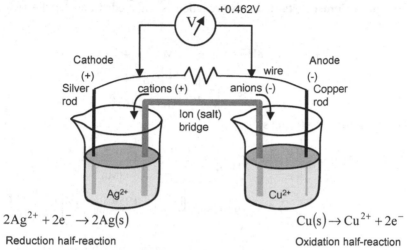

$$2Ag^{2+} + 2e^- \rightarrow 2Ag(s)$$
Reduction half-reaction

$$Cu(s) \rightarrow Cu^{2+} + 2e^-$$
Oxidation half-reaction

How long does the spontaneous reaction proceed for? Does this cell "run down"? Does the voltmeter reading start to drop and then go to zero?

Initially, we know from our standard electrode potentials that the attainment of equilibrium, the objective of all chemical reactions, favours the forward reaction as written:

$$Cu(s) + 2Ag^{2+}(aq) \rightarrow Cu^{2+}(aq) + 2Ag(s)$$

As the reaction proceeds, the Ag^{2+} ions are being used up in the cathode solution, and there is a build-up of Cu^{2+} ions in the anode solution. Electrons lose potential energy as they pass through the wire and this energy is converted to heat in the resistor. When the electrons combine with the silver ions at the cathode, they are in a lower energy state compared to when they were bound to the copper. Attainment of a lower energy state is the driving force behind most spontaneous chemical reactions.

When the silver ions are in short supply, then not all the electrons coming from the wire will find a home to go to and there will be a slight increase in negative charge at the cathode. When there are no more silver ions to accept these electrons, the build-up of negative electric charge will be great enough to prevent any more from arriving. At this point, the electrical potential difference between the anode and cathode will be zero and the cell will be exhausted.

10.13 Equilibrium Constant

The potential of an electrochemical cell depends upon the concentration of the reactants. When the cell is exhausted, we reach a state of equilibrium and the **equilibrium constant** for the reaction is satisfied. Consider the redox reaction:

$$5Fe^{2+} + MnO_4^- + 8H^+ \rightarrow 5Fe^{3+} + Mn^{2+} + 4H_2O$$

$$K_c = \frac{\left[Fe^{3+}\right]^5 \left[Mn^{2+}\right]}{\left[Fe^{2+}\right]^5 \left[MnO_4^-\right]\left[H^+\right]^8}$$

At equilibrium, $E = 0$, and so

$$E_{Fe}^o + 0.059 \log_{10} \frac{\left[Fe^{3+}\right]}{\left[Fe^{2+}\right]} = E_{MnO_4^-}^o + \frac{0.059}{5} \log_{10} \frac{\left[MnO_4^-\right]\left[H^+\right]^8}{\left[Mn^{2+}\right]}$$

$$E_{MnO_4}^o - E_{Fe}^o = 0.059 \log_{10} \frac{\left[Fe^{3+}\right]}{\left[Fe^{2+}\right]} - \frac{0.059}{5} \log_{10} \frac{\left[MnO_4^-\right]\left[H^+\right]^8}{\left[Mn^{2+}\right]}$$

$$\Delta E^o = \frac{0.059}{5} \left(\log_{10} \frac{\left[Fe^{3+}\right]^5}{\left[Fe^{2+}\right]^5} - og_{10} \frac{\left[MnO_4^-\right]\left[H^+\right]^8}{\left[Mn^{2+}\right]} \right)$$

$$= \frac{0.059}{5} \log 10 \frac{\left[Fe^{3+}\right]^5 \left[Mn^{2+}\right]}{\left[Fe^{2+}\right]^5 \left[MnO_4^-\right]\left[H^+\right]^8}$$

$$= \frac{0.059}{5} \log_{10} K_c$$

$$K_c = \log 10 \frac{5\Delta E^o}{0.059} \qquad\qquad E^o{}_{MnO_4^-} = +1.51V$$

$$= 5.1 \times 10^{62} \qquad\qquad\qquad E^o{}_{Fe} = +0.77V$$

10.14 Lead/Acid Battery

A **battery** is a collection of cells arranged in series. In a motor vehicle **lead acid battery**, there are six cells, each producing about 2V, arranged in series to provide 12V for the vehicle. The redox couples involved are:

$$PbSO_4(s)/Pb(s), SO_4^{2-} \qquad PbO_2(s), H^+, SO_4^{2-}/PbSO_4(s)$$
$$E^\circ = -0.36V \qquad\qquad E^\circ = +1.69V$$

Since the standard potential for the PbO_2 reaction is higher, then this species undergoes reduction and is the cathode.

Reduction (cathode) $PbO_2(s) + 4H^+ + SO_4^{2-} + 2e^- \rightarrow PbSO_4(s) + 2H_2O$

Oxidation (anode) $Pb(s) + SO_4^{2-} \rightarrow PbSO_4(s) + 2e^-$

$PbO_2(s) + 4H^+ + 2SO_4^{2-} \rightarrow 2PbSO_4(s) + 2H_2O$ **Discharge** reaction

$$E^\circ_{cell} = E^\circ_{cathode} - E^\circ_{anode}$$
$$= 1.69 - -0.36$$
$$= 2.05V$$

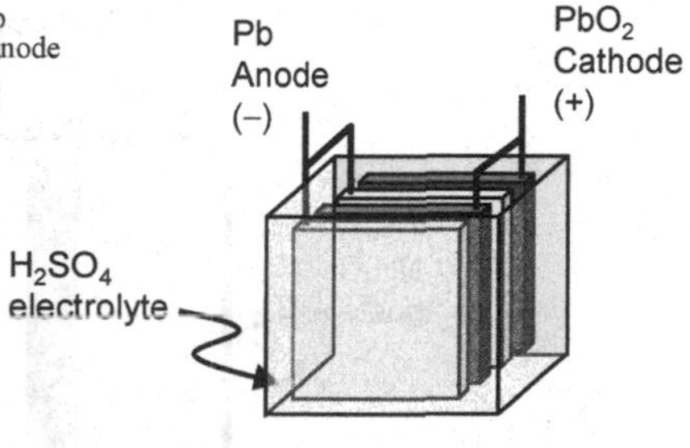

Pb
Anode
(–)

PbO$_2$
Cathode
(+)

H$_2$SO$_4$
electrolyte

The concentration of H_2SO_4 in a motor vehicle battery is usually about 5M and so the actual cell potential is about 2.15V.

The **anode** (the negative plates) is solid lead, Pb, while the **cathode** (the positive plates) is **lead peroxide**, PbO_2, in a solution of **sulphuric acid**. The reaction product, **lead sulphate** $PbSO_4$, sometimes appears as a white precipitate on the battery terminals.

As the cell discharges, the concentration of H_2SO_4 decreases and the density of the solution also decreases. The density of the solution is thus a measure of the state of **charge** of the battery and is measured with a **hydrometer**.

When the battery is charged, the above reaction proceeds in reverse. Overcharging results in the evolution of gaseous H_2 at the anode and O_2 at the cathode, and the water in the electrolyte has to be replenished.

10.15 Dry Cell

In a conventional **dry cell**, or **Leclanché cell**, oxidation occurs at a zinc anode and reduction occurs at an inert carbon cathode. The electrolyte is not a liquid, but a paste of magnesium dioxide MnO_2, zinc chloride $ZnCl_2$, ammonium chloride NH_4Cl, and carbon black particles.

$$Zn(s) \rightarrow Zn^{2+}(aq) + 2e^-$$ Oxidation (anode −)

$$2MnO_2(s) + H_2O + 2e^- \rightarrow Mn_2O_3(s) + 2OH^-$$ Reduction (cathode +)

$$NH_4^+(aq) + OH^-(aq) \rightarrow NH_3(g) + H_2O$$ Acid/base absorbs the

$$Zn^{2+}(aq) + 4NH_3(aq) \rightarrow [Zn(NH_3)_4]^{2+}(aq)$$

Acid/base absorbs the OH⁻ ions
Complex ion formation absorbs NH_3 gas

$$Zn(s) + 2MnO_2(s) \rightarrow ZnMn_2O_4(s)$$ Overall cell reaction

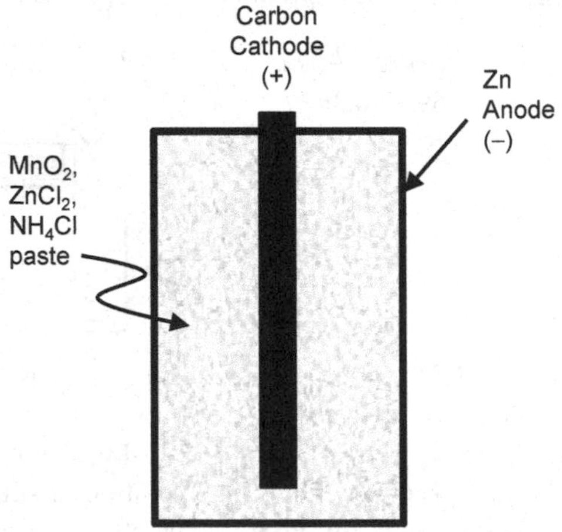

Carbon
Cathode
(+)

Zn
Anode
(−)

MnO_2,
$ZnCl_2$,
NH_4Cl
paste

Modern **alkaline cells** use an alkaline electrolyte of KOH instead of the MnO_2, $ZnCl_2$, NH_4Cl paste of the zinc-carbon type of dry cell. In an alkaline cell, the half-reactions are

$$Zn(s) + 2OH^-(aq) \rightarrow ZnO(s) + H_2O(l) + 2e^-$$

$$2MnO_2(s) + H_2O(l) + 2e^- \rightarrow Mn_2O_3(s) + 2OH^-(aq)$$

10.16 Corrosion

Corrosion is the undesirable oxidation of a metal. The most well-known example is the **rusting of iron**. In simple terms, the complex redox reaction for the rusting of iron is written as

$$4Fe(s) + 3O_2(g) \rightarrow 2Fe_2O_3(s)$$

In practice, the rusting of iron requires both air (O_2) and water, and the reaction proceeds as a series of steps.

$$Fe(s) \rightarrow Fe_2^{2+} + 2e^-$$

Oxidation half-reaction

$$O_2 + 2H_2O + 4e^- \rightarrow 4OH^-$$

Reduction half-reaction

The OH⁻ ions react with the Fe^{2+} ions:

$$Fe^{2+} + 2OH^- \rightarrow Fe(OH)_2$$

then $4Fe(OH)_2 + O_2 + 2H_2O \rightarrow Fe(OH)_3(s)$

and finally $Fe(OH)_3(s) \rightarrow Fe_2O_3, H_2O + 2H_2O$

The brown coloured metal we usually associate with rust is the compound Fe_2O_3, H_2O, an Fe(III) hydrated oxide

Iron and steel often rust badly because the rust has a lower density than the parent metal and tends to flake off, thus exposing more of the material to water and oxygen and promoting further rusting.

Water and oxygen are typically excluded by painting the iron article. However, a very effective way to reduce rusting of iron is to attach a more reactive metal (i.e. one with a lower $E°$ so that the iron undergoes reduction and the **sacrificial anode** (usually zinc) undergoes oxidation. This process is often called **cathodic protection**.

Fe / Fe_2^{2+}	Zn^{2+} / Zn	Iron or steel coated with zinc is said to be **galvanised** (since it acts like a galvanic cell).
$E° = -0.44V$	$E° = -0.76V$	

Water and oxygen are typically excluded by painting the iron article. However, a very effective way to reduce rusting of iron is to attach a more reactive metal (i.e. one with a lower $E°$ so that the iron undergoes reduction and the sacrificial metal (usually zinc) undergoes oxidation.

Aluminium undergoes corrosion in the presence of atmospheric oxygen. A very thin (nm) tough layer of Al_2O_3 (**alumina**) appears within nanoseconds on the metallic surface which shields the parent metal from further corrosion. Alumina is a very hard ceramic material and is a non-conductor of electricity.

10.17 Electrolysis

An **electrolytic cell** is not a spontaneous redox reaction, and requires an electrical energy input to proceed.

Consider the redox reaction:

$$Cu^{2+}(aq) + 2Ag(s) \rightarrow Cu(s) + 2Ag^{2+}(aq)$$

Arranged in this way, the half-reactions are:

$$2Ag(s) \rightarrow 2Ag^{2+}(aq) + 2e^- \qquad Cu^{2+}(aq) + 2e^- \rightarrow Cu(s)$$

Oxidation half-reaction   Reduction half-reaction

-0.799V   +0.337V

The total potential is −0.462V and so this is not a spontaneous reaction. If a cell were constructed and connected with a wire between the two electrodes, the reaction would proceed in the reverse to that as shown above. However, if we apply an external **potential difference** to the electrodes, say a 1V potential, then

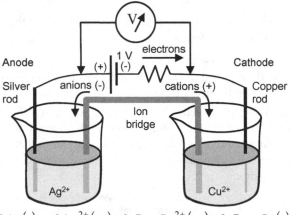

Arranged in this way, we have an **electrolytic cell** in which the copper rod is the cathode (+) and the silver rod is the anode (-). The reaction proceeds as written.

$$2Ag(s) \rightarrow 2Ag^{2+}(aq) + 2e^- \qquad Cu^{2+}(aq) + 2e^- \rightarrow Cu(s)$$

Oxidation half-reaction   Reduction half-reaction

Note: **oxidation** occurs at the **anode** which in this case is the silver. **Reduction** occurs at the **cathode**, in this case the copper. The silver rod dissolves away and copper particles coat the copper rod.

The applied external potential has to overcome +0.462V in order for the reaction to proceed, and so the potential difference across the electrodes would be 1 − 0.462 = 0.538V.

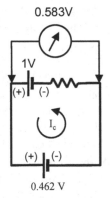

I_c indicates conventional current flow

11. Carbon Chemistry

Summary

Functional groups

H	Alkanes	Single C bond
C=C	Alkenes	Double C bond
C≡C	Alkynes	Triple C bond
OH	Alcohols	ROH
O	Ethers	ROR'
CHO	Aldehydes	RCHO
CO	Ketones	R(C=)OR
COOH	Acids	RCOOH
COO	Esters	RCOOR'
NH$_2$	Amines	RNH$_2$ R$_2$NH, R$_3$N

Alkyl groups

Methyl
Ethyl
Propyl
Butyl
Pentyl
Hexyl
Heptyl
Octyl
Nonyl
Decyl

 Benzene ring

11.1 Carbon

The element **carbon** has four outer valence electrons and is the first element of Group IV in the periodic table. The other elements in this group become more metallic in nature with increasing atomic number. At room temperature, carbon is relatively inert. At high temperatures, it can form covalent bonds with itself and many other elements. Carbon compounds are the basis for all life on Earth. It is interesting to note that **silicon**, the next element down in the group and also with four valence electrons, is the basis of the composition for nearly all of the Earth's minerals.

With four valence electrons, carbon has the choice of either losing or gaining four electrons to achieve a noble gas configuration but in practice, it tends to share its four electrons via covalent bonds. On the other hand, tin and lead, the heaviest elements in Group IV, with more loosely bound valence electrons, find it energetically favourable to share these electrons in the form of **metallic bonds**.

Carbon reacts with metals to form carbides: e.g. silicon carbide SiC; with non-metals to form molecular compounds: e.g. methane CH_4; with oxygen to form carbon monoxide CO and carbon dioxide CO_2; and with nitrogen to form cyanides (e.g. HCN).

Pure carbon forms three crystalline structures found naturally: **diamond, graphite,** and the **fullerenes**. These are **allotropes**. These have significantly different mechanical, thermal and electrical properties and differ only in the geometrical arrangement of their atoms in the crystal structures.

Diamond has a three-dimensional tetrahedral structure. Each carbon atom has four neighbours, and all the valence electrons are equally shared with no vacancies or lone electrons. Thus, diamond is a poor electrical conductor, has a high **hardness** and **elastic modulus**. Diamond is a good heat conductor due to the free passage of **phonons**.

Graphite has a two-dimensional hexagonal sheet structure with the sheets bonded together as layers by **van der Waals forces**. Each carbon atom has three neighbours. Thus, three valence electrons are involved in covalent bonding while the fourth is left in a relatively free state and is responsible for graphite being an electrical conductor.

The **fullerene** form of carbon consists of a three-dimensional arrangement of C atoms in a pentagon which together form a sphere or tube. The most famous is the C_{60} molecule, often referred to as a **buckyball**.

11.2 Carbon Compounds

Carbon forms a great number of compounds with hydrogen. These are called **hydrocarbons**. In addition, **hydrocarbon derivatives** are formed when other atoms in addition to hydrogen become involved in the bonding.

Central to this remarkable ability to form such a wide variety of compounds is the formation of chains of atoms. For example, consider the simple hydrocarbon **methane** CH_4:

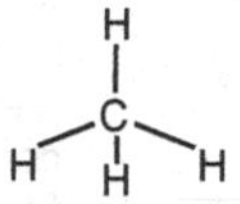

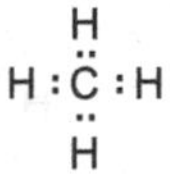

This molecule can be easily expanded to form **ethane** C_2H_6:

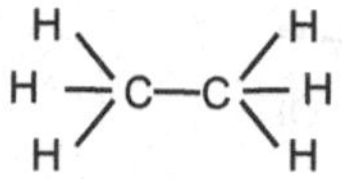

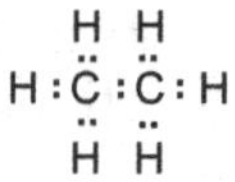

And so on for **propane**, C_3H_8, and **butane**, C_4H_{10}.

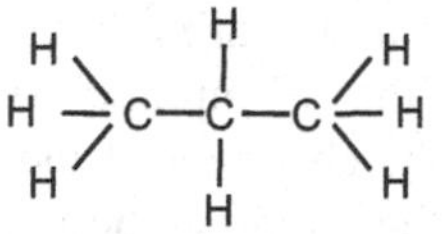

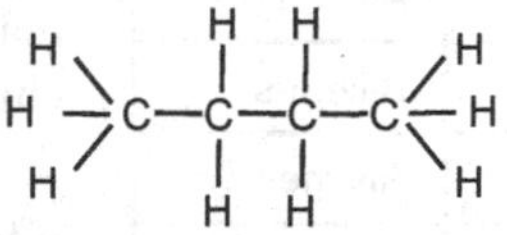

Other possibilities abound, including compounds involving a double carbon bond such as in **ethylene** and the triple bond in **acetylene**:

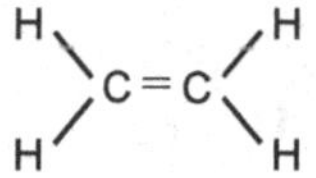

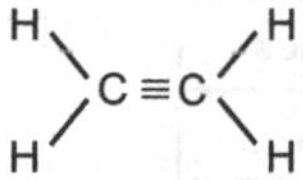

Compounds with single bonds between carbon atoms are **saturated hydrocarbons**. Compounds with double or triple bonds are called **unsaturated hydrocarbons**. Hydrocarbons with the same chemical formula but with different structures are possible. For example, **butene** C_4H_8, can exist in four different structural arrangements, each of which is called an **isomer**.

The reason for this is that saturated compounds are saturated with H atoms. When there is a double bond, the maximum possible number of H atoms that can attached to a C atom is decreased and so the C atoms is unsaturated.

The substitution of other elements in place of one or more hydrogen atoms in carbon compounds is responsible for the existence of a large number of hydrocarbon derivatives. Alcohols are hydrocarbons with an attached OH **functional group**.

The ability of carbon to form chains, double bonds, isomers and contain functional groups is responsible for the great variety of carbon compounds which form the chemistry of life on Earth – or **organic chemistry**.

11.3 Functional Groups

Hydrocarbon derivatives are formed when other atoms in addition to hydrogen become involved in the bonding. In these cases, the core hydrocarbon is written with an attached **functional group**. The core or hydrocarbon compound (also called the **remainder** or **residue**) is written with the symbol R. A common functional group is the alcohol, or OH group.

methanol CH_3OH:

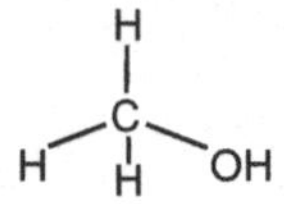

Alcohols can be written as ROH where R is the hydrocarbon residue, or the core hydrocarbon component, and OH is the alcohol functional group.

ethanol C_2H_5OH:

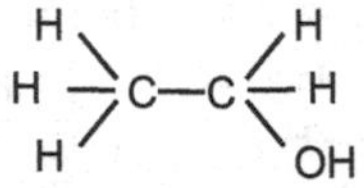

Functional groups

H	Alkanes
C=C	Alkenes
C≡C	Alkynes
OH	Alcohols
O	Ethers
CHO	Aldehydes
CO	Ketones
COOH	Acids
COO	Esters
NH₂	Amines

Functional groups determine the nature of the reactivity of the hydrocarbon involved. The chemistry of the compound is usually determined by the character of the functional group while the hydrocarbon residue is relatively inert.

Consider some compounds based upon the $CH_3CH_2 -$ residue:

Ethane: $CH_3CH_2 - H$

Ethanol: $CH_3CH_2 - OH$

Ethyl bromide: $CH_3CH_2 - Br$

The CH_3CH_2- residue is called the **ethyl group**. Similarly, the CH_3- is the **methyl group** and the $CH_3CH_2CH_2-$ is the **propyl group**.

These named **alkyl groups** form bonds with various different functional groups and the resulting compounds R–X are thus **derivatives** of the parent hydrocarbon (e.g. methane, ethane, etc).

11.4 Alkanes

Alkanes are **saturated hydrocarbons** since the bonds with the carbon atoms are single bonds and they contain only carbon and hydrogen. The names of the compounds are derived from the number of carbon atoms.

The first 10 alkanes

Methane	CH_4
Ethane	C_2H_6
Propane	C_3H_8
Butane	C_4H_{10}
Pentane	C_5H_{12}
Hexane	C_6H_{14}
Heptane	C_7H_{16}
Octane	C_8H_{18}
Nonane	C_9H_{20}
Decane	$C_{10}H_{22}$

Alkanes can have chain or branched chain (**aliphatic**), or cyclic (**alicyclic**) structures. As the complexity of the molecules increases, there are increased ways in which molecules can form structures with the same molecular formula.

Compounds having the same molecular formula but different structure are called **isomers**.

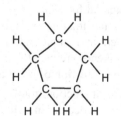

n-pentane (or normal pentane):

Alkanes are the principal components of **petroleum**. Low molecular weight compounds are generally gases at room temperature and pressure. Gasoline (petrol) and turpentine typically contain light fractions (C_7H_{16} to $C_{10}H_{22}$). Kerosene, diesel oil, lubricating oils, pitch, and so on consist of heavier compounds. Alkanes are relatively inert, but can undergo exothermic oxidation (**combustion**) which make them useful sources of energy.

Alkanes with a chain structure have the molecular formula C_nH_{2n+2}.

iso-pentane (single branch at the end)

Alkanes having a cyclic or a ring structure have the formula C_nH_{2n}.

cyclo-pentane (ring structure)

Fuel	Calorific value (MJ/kg)
Petrol	45
Diesel	44.4
Ethanol	27.2

Alkanes are sometimes called **paraffins** because they have little (parum) affinity for forming bonds with other atoms or molecules. When they do participate in reactions, this usually involves the **substitution** of a hydrogen atom with another atomic species.

11.5 Alkenes

Alkenes (also known as **olefins**) are unsaturated hydrocarbons because they contain a double bond between two of the available carbon atoms instead of single bonds. As a consequence, there are fewer than four atoms attached to each of the carbon atoms.

Alkenes in a chain configuration have the general formula C_nH_{2n}. Ethene or **ethylene** (C_2H_4) is the simplest alkene.

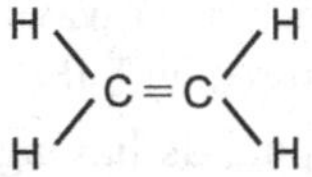

Bond length: 1.33 Å
Energy: 606 kJ/mol
ΔH = +52.5 kJ/mol

Note that the carbon double bond can occur in different places in the structure, leading to formation of isomers. In **propylene** (C_3H_6):

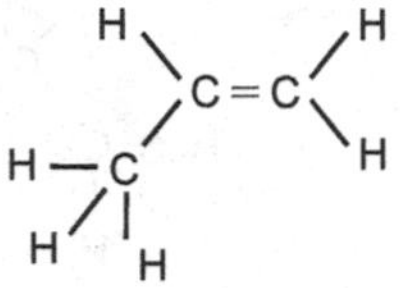

Ethene, propene, and butene are gases. Higher alkenes are generally liquids and those with greater than sixteen carbon atoms are generally waxy solids.

While alkenes are relatively inert, they are more reactive than alkanes because of the double bond. The carbon double bond has a higher bond strength compared to the single bond, but the double bond allows a greater potential for attachment for other atoms and this leads to greater reactivity compared to single-bonded structures. Chemical reactions involving alkenes are usually associated with the breaking of the double bond and the **addition** of other atoms.

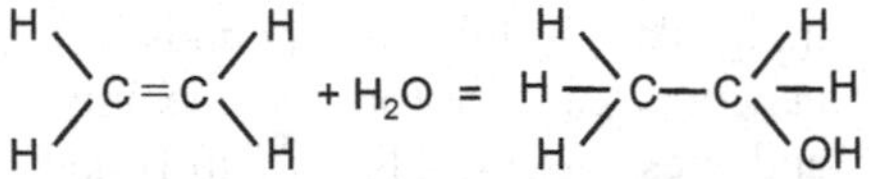

ethylene ethanol

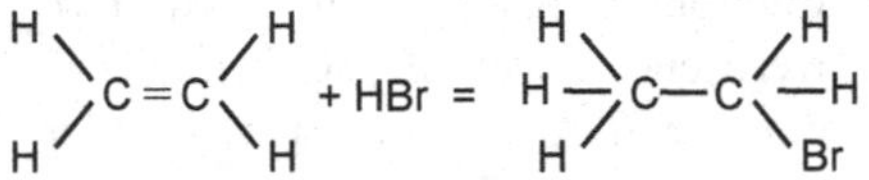

ethylene bromoethane

An important reaction involving alkenes is that of **polymerisation**, where a chain reaction occurs in which the C=C double bonds are successively broken and employed to form larger and larger single-bonded structures.

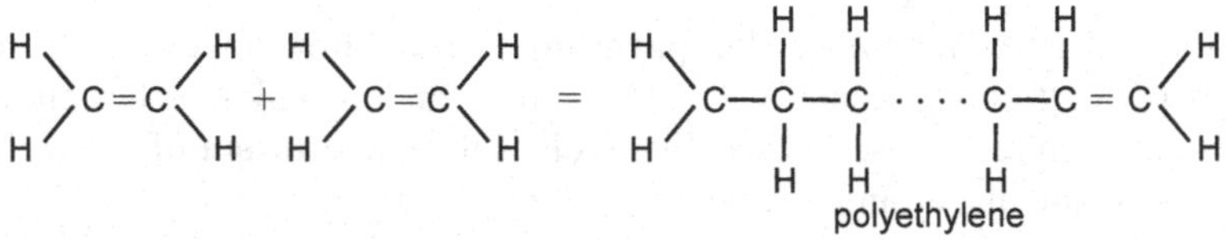

polyethylene

11.6 Alkynes

Alkynes are similar to alkenes except that they feature a carbon-carbon triple bond. Alkynes are unsaturated hydrocarbons and have the general formula C_nH_{2n-2}. Alkynes are also known as **acetylenes**, but the common use of this term refers to **ethyne**: C_2H_2.

C_2H_2 $H-C\equiv C-H$
acetylene

Bond length:	1.2 Å
Bond energy:	828 kJ/mol
$\Delta H =$	+226.9 kJ/mol

The triple bond can occur anywhere in the chain in higher alhynes. Consider the compound 1-butyne. The 1- indicates that the triple bond is at the end of the chain: $CH_3CH_2C\equiv CH$

The CH_3CH_2 is the ethyl group, and so this compound can also be named **ethylacetylene**.

Alkynes have a similar chemistry to alkenes. The $C\equiv CH$ unit is a linear structure and in acetylene, the molecules are therefore rod-like in shape.

The triple bond is capable of opening up and accommodating additional atoms during reactions. Additions may occur with halides, halogens, water, and ammonia.

Ethyne	$HC\equiv CH$
Propyne	$CH_3C\equiv CH$
1-Butyne	$CH_3CH_2C\equiv CH$
1-Pentyne	$CH_3(CH_2)_2C\equiv CH$
1-Hexyne	$CH_3(CH_2)_3C\equiv CH$
1-Heptyne	$CH_3(CH_2)_4C\equiv CH$
1-Octyne	$CH_3(CH_2)_5C\equiv CH$
1-Nonyne	$CH_3(CH_2)_6C\equiv CH$
1-Decyne	$CH_3(CH_2)_7C\equiv CH$

The $C\equiv CH$ unit, located at the end of the structure, imparts an acidic nature to alkynes and so acid-base reactions are also possible. Alkynes can also undergo **polymerisation** to give **polyacetylenes**.

One of the most common uses of acetylene is in gas welding. The combustion of acetylene in oxygen provides a flame temperature of about 3300°C making it one of the hottest flames available in industry.

$$2C_2H_2 + 5O_2 \rightarrow 4CO_2 + 2H_2O$$

Acetylene is a gas, but an unstable one, especially at moderate to high pressures. In welding gas bottle sets, acetylene is usually dissolved in liquid **acetone** or dimethylformamide and stored under pressure in a cylinder containing a porous medium, **agamassam**.

11.7 Benzene

An important class of carbon compounds arises when carbon atoms are arranged in a special ring formation.

The **benzene ring** C_6H_6 has all the atoms in a plane, with a 120° angle between the carbon atoms. The initial difficulty with this structure is that each carbon atoms has three neighbours sharing one electron, but there are four valence electrons in carbon. Evidently, there either has to be three double bonds present or some way of accommodating the fourth valence electron. If each second carbon atom had a double bond, then all available valence electrons would be shared, but experimental evidence indicates that the bond strength for the C-C bonds are equal around the ring, and that it is stronger than a single bond, but not as strong as a double bond.

The bonding in a benzene ring is explained by the allowable movement of the **valence electrons** in the ring structure. Rather than forming single or double bonds, electrons in the ring form a resonance hybrid,   or a superposition of single and double bonds,   whereby they may have movement throughout the ring, much like valence electrons in a metal have movement throughout the **conduction band**.

To signify this, the benzene ring is often written in shorthand form using one of the following diagrams:

It should be always remembered that there are not single or double bonds, but a special bond somewhere between the two based upon electron movement around the ring.

Ring compounds of this type are called **aromatic compounds**. When they consist of C and H only, they are called **aromatic hydrocarbons**. Note: **aliphatic** compounds in a ring structure are not aromatic; they are simple ring structures (**alicyclic**). Aromatic compounds have mobile electrons within the ring and it is these that determine the chemistry of these compounds. In contrast to the addition reactions for alkenes with a chain structure,   benzene undergoes substitution reactions in which H atoms are replaced by other atoms.

11.8 Alcohols

Hydrocarbon derivatives arise when functional groups combine with a residual hydrocarbon core. **Alcohols** are those compounds which have one or more OH (**hydroxyl**) groups attached, ROH.

Three types of alcohol are to be found depending on the number of C atoms attached to the C atom with the OH group. Primary (n) alcohols are written RCH_2OH.

$$CH_3CH_2CH_2 - OH \quad \begin{array}{l} \text{1-propyl alcohol} \\ \text{n-propanol} \end{array}$$

Secondary (s) alcohols are written RR'CHOH:

$$\begin{array}{cc} OH & \text{isopropyl alcohol} \\ | & \text{2-propanol} \\ CH_3CHCH_3 & \text{s-propanol} \end{array}$$

Tertiary (t) alcohols are RR'R"COH.

$$\begin{array}{cc} OH & \\ | & \\ CH_3 - C - OH & \text{t-butyl alcohol} \\ | & \\ CH_3 & \end{array}$$

Lower alcohols are soluble in water due to the presence of the polar OH group. In higher alcohols, the longer the carbon chain, the less water-like the compound is and behaves more like a hydrocarbon – volatile and relatively inert. **Pentanol** and higher compounds are insoluble in water.

Alcohols can react with other atoms and can act as weak **bases** (due to the unshared electron pairs on the O atom) or weak **acids** – where the H atom on the OH group can act as a proton donor.

$$-C - \overset{H}{\overset{\diagup}{O}}:$$

Phenol (hydroxybenzene) has the OH group attached to a benzene ring and forms the basis of synthesis for many other compounds.

Ethanol (C_2H_5OH) is "alcohol" to the common person (e.g. in alcoholic drinks). **Denatured ethanol** is the usual form used for cleaning and other uses. Additives render it unpalatable and, in some cases poisonous (such as when methanol is added to the ethanol to give **methylated spirits**).

Methanol (CH_3OH) is highly toxic and may cause blindness or death when only a few millilitres are ingested.

A common cleaning alcohol is **isopropyl alcohol**, or 2-propanol ($CH_3)_2CHOH$.

Ethylene glycol $C_2H_4(OH)_2$ is commonly used as an anti-freeze/ anti-boil additive in motor vehicle cooling systems. In its pure form it has a boiling point of 197.3 °C but has a lower heat capacity than water and so is used diluted. It does this by depressing the hydrogen bonding between water molecules so that, upon freezing, the temperature must be significantly lowered before ice crystals can form.

An interesting compound prepared from phenol, sodium hydroxide, and carbon dioxide is **aspirin**.

11.9 Ethers

Ethers have the general formula ROR' where R is an alkyl group. They can occur as aliphatic (chained), alicyclic, or aromatic (**benzene ring**) formations, or a combination. They are relatively unreactive hydrocarbon derivatives.

The lone pairs of electrons on the O atom allow ethers to form **hydrogen bonds** with water molecules and so they are generally partially soluble in water.

$$CH_3CH_2 \overset{\overset{..}{\overset{..}{O}}}{\diagup} \diagdown CH_3CH_2$$
diethyl ether

diphenyl ether

Perhaps one of the most well known ethers is **diethyl ether**, $CH_3CH_2OCH_2CH_3$, a low boiling point colourless liquid used as an anaesthetic. Diethyl ether can be formed from the removal of water from ethanol.

However, they cannot form hydrogen bonds between themselves and so generally have a lower boiling point than comparable alcohols.

In some respects, ethers are similar to alcohols in that they can be considered derivatives of water. In an alcohol, the R takes the place of one of the H atoms in the water molecule. In an ether, both of the H atoms are replaced by alkyl groups R and R'. Ethers are generally isomers of the corresponding alcohols. For example, ethanol, C_2H_5OH, has the same formula as dimethylether, but of course has quite different chemical and physical properties.

Dimethyl ether, CH_3OCH_3, (or **DME**) is a gas at room temperature is often used as an aerosol propellant since it can be stored as a liquid under pressure in equilibrium with the gas phase, evaporates readily and leaves the desired product as an aerosol mist for the application.

Methyl phenyl ether $CH_3OC_6H_5$ is found in aniseed oil.

Ethers are generally unreactive substances and can withstand attack from either acids or bases, but are more reactive than comparable alkanes.

Simple ethers have the alkyl groups on each end of the oxygen atom and if the groups are the same, are thus named dialkyl ether (such as in dimethylether). If one of the groups is a more complex group, then the structure of the name is alk-oxy-alkane. For example, $CH_3CH_2CH_2OCH_3$ is **methoxypropane**.

11.10 Aldehydes

Alcohols can be thought of as a step towards complete oxidation of the corresponding alkane. Primary alcohols can also undergo oxidation to form **aldehydes** and **carboxylic acids**, and secondary alcohols can oxidise to form **ketones**.

Aldehydes have a carbon-to-oxygen double bond (which itself is called the **carbonyl group**) with the general formula RCHO. They are usually prepared by **oxidation** of the corresponding alcohol.

Perhaps one of the most well known aldehyde is that made from **formaldehyde** dissolved in water to give **formalin**, a preservative fluid used for embalming biological tissues.

aldehyde group

Consider the reaction between **methanol** and potassium dichromate:

$$3CH_3OH + Cr_2O_7^{2-} + 8H^+ \rightarrow 3CH_2O + 2Cr^{3+} + 7H_2O$$

In this equation, the Cr_2 ion has been reduced to the Cr^{3+} ion, and the methanol has lost two electrons (and two protons H^+) and hence has been oxidised to give CH_2O, formaldehyde.

methanal (formaldehyde)

Aldehydes often have a pleasant odour or flavour, such as in the benzene derivative vanillin:

Methanal	HCHO	Formaldehyde
Ethanal	CH$_3$CHO	Acetaldehyde
Propanal	CH$_3$CH$_2$CHO	Propionaldehyde
Butanal	CH$_3$CH$_2$CH$_2$CHO	Butyraldehyde

To avoid confusion with alcohols (OH), the carbonyl group with an attached H is written CHO and not COH.

Cyclic forms also exist, such as **benzaldehyde**, an aromatic aldehyde:

and cinnamaldehyde, the active ingredient in cinnamon.

Aldehydes are closely related to **ketones**, but have the **carbonyl group** C=O at the end of a carbon chain rather than in between two C atoms. Lower aldehydes are soluble in water. Aldehydes can be further oxidised to form **carboxylic acids**.

11.11 Ketones

Ketones have a **carbonyl group** C=O attached to two other C atoms that form part of the same or different **alkyl groups**. (In an aldehyde, one of the other atoms is an H atom.) The general formula is RC(=O)R'.

The oxygen atom is more **electronegative** than attached carbon (i.e. the electron pairs that constitute the double bond are closer to the oxygen atom than the carbon atom), making this group polar. Unlike aldehydes, ketones do not have an H atom attached to the carbonyl group.

$$CH_3 - C \overset{\displaystyle O}{\underset{\displaystyle CH_3}{\Big\backslash}}$$

2-propanone
(acetone)

Acetone can be formed by the oxidation of 2-propanol according to

$$3CH_3CHOHCH_3 + Cr_2O_7^{2-} + 8H^+ \rightarrow$$

$$3CH_3COCH_3 + 2Cr^{3+} + 7H_2O$$

Aldehydes and **ketones** have many similarities and both are formed from the oxidation of an alcohol.

Ketones cannot be further oxidised to an acid since there is no H atom available on the functional group (as in the case of **aldehydes**). The absence of an H atom bonded to an oxygen atom means that both aldehydes and ketones do not form **hydrogen bonds** with each other and so generally have a lower boiling point than comparable alcohols. However, the lone pairs of electrons on the oxygen atom can form a hydrogen bond with an H atom in a water molecule and so lower ketones (and also aldehydes) are generally soluble in water.

The names of ketones (and aldehydes) are formed by the number of C atoms in the longest chain, and include the C atom in the carbonyl group.

Acetone CH_3COCH_3 is widely used as a relatively non-toxic solvent for various carbon compounds (e.g. paint, plastics, nail varnish, superglue, etc). It also has the useful property of being soluble in water and is commonly used as a drying agent for glassware in chemistry laboratories. It is very volatile and dries without leaving a residue.

Another common ketone used as a solvent for paints, varnish and gums is 2-butanone, or **methyl ethyl ketone** (**MEK**), or methyl acetone, $CH_3COCH_2CH_3$, a flammable clear liquid that is soluble in water. It has a characteristic sweet odour. As well as industrial uses as a solvent, it is also used as a polystyrene cement in plastic model kits.

Another significant ketone, with a cyclic (but not aromatic) structure, is **cyclohexanone** $(CH_2)_5CO$ which is used in the manufacture of **nylon**.

cyclohexanone

Ketones have significant biochemical functions and are produced when body fat is used as a source of energy when there is a shortage of **glucose** (such as during fasting).

11.12 Carboxylic Acids

Hydrocarbon derivatives that contain both the carbonyl (C=O) and hydroxyl (OH) functional groups are called **carboxylic acids** with the general formula RCOOH. When R is an aliphatic residue, the compound is called a **fatty acid**. When R is an aromatic ring, the acid is a **benzene ring derivative**.

The COOH group is called the **carboxyl group**. Carboxylic acids can be prepared from the oxidation of the corresponding alcohol.

ethanoic acid
(acetic acid)

benzoic acid

A very common aliphatic acid is **acetic acid**, written CH_3COOH and is found in vinegar. Formic acid (from the latin formica for ant) in the found in venom of ants. Stearic acid is found in soaps, waxes and plant oils. Other carboxylic acids can be found in coconuts, chocolate,

$$5CH_3OH + 4MnO_4^- + 12H^+ \rightarrow$$

$$5HCOOH + 4Mn^{2+} + 11H_2O$$

In this reaction CH_3OH and H_2O produces HCOOH and four protons and four electrons. Since there is a loss of four electrons (compared to two for the oxidation of methanol to methanal, or formaldehyde), formic acid is a more highly oxidised compound than **formaldehyde**.

Some carboxylic acids:

Formic	Methanoic	HCOOH
Acetic	Ethanoic	CH_3COOH
Propionic	Propionic	CH_3CH_2COOH
Butyric	Butanoic	CH_32CH_2COOH
Stearic	Octa decanoic	CH_316CH_2COOH

Similarly, ethanol can be oxidised to form **acetic acid**. Acetic acid can also be formed from the oxidation of acetaldehyde. Further oxidation of these compounds can be obtained via the rather energetic process of **combustion**, in which case the product is CO_2 and water.

Carboxylic acids are weak acids and dissociate only slightly, giving H^+ ions in water, and act as proton donors in reactions. The H^+ ions come from the COOH functional group. Lower carboxylic acids are soluble in water.

A common chemical demonstration reaction involves that between vinegar and **baking soda** to give sodium ethanoate, carbon dioxide and water:

$$CH_3COOH + NaHCO_3 \rightarrow 3CH_3COONa + CO_2 + 7H_2O$$

There is an ionic bond between the CH_3COO^- and Na^+ ions. By convention, the ionic compound is written with the metal ion last.

Carboxylic acids react with alcohols to form fats and oils (esters). **Fatty acids** are those needed by the body and are obtained from digestion of animal and vegetable oils and fats.

11.13 Esters

Carboxylic acids contain the functional group COOH. The H atom can be easily given up (proton donor): hence the acid classification.

Derivatives of carboxylic acids are formed when the OH group is replaced by something else. When the OH group is replaced by OR, then an **ester** is formed. Esters are generally colourless liquids insoluble in water and have the general formula RCOOR'. For example, a reaction between methanol and acetic acid forms the ester **methyl acetate**:

$$CH_3OH + CH_3COOH \rightarrow CH_3COOCH_3 + H_2O$$

Note that the methyl group is the CH_3 at the end of the molecule. The "acetate" is from the carboxylic acid. Esters are written with the functional group first.

$$CH_3COO - C \overset{O}{\underset{O - CH_3}{\diagdown}}$$

methyl acetate

In this reaction, water is formed from the H atom in the incoming alcohol, and the OH group from the carboxylic acid, leaving behind an ester.

Esters are commonly used in food flavourings and perfumes due to their pleasant odour.

Methyl butyrate	Apple	$CH_3CH_2CH_2COOCH_3$
Ethyl butyrate	Strawberry	$CH_3CH_2CH_2COOCH_2CH_3$
Pentyl butyrate	Apricot	$CH_3CH_2CH_2COO(CH_2)_4CH_3$
Pentyl acetate	Banana	$CH_3COO(CH_2)_4CH_3$
Isoamyl acetate	Pear	$CH_3COOCH_2CH_2\ CH(CH_3)_2$
Octyl acetate	Orange	$CH_3COO(CH_2)_7CH_3$

Most naturally occurring fats and oils are very large esters made from organic acids and alcohols (usually **glycerol**). When there are no C=C bonds, the acid is said to be **saturated**. When the acid contains C=C bonds, it is **unsaturated**. When an acid has one C=C bond it is **monounsaturated**, and when there is more than one C=C bond, it is referred to as being **polyunsaturated**. These **fatty acids** form saturated, unsaturated and polyunsaturated fats and oils (oils are esters that are liquid at room temperature whereas waxes are solid).

Esters also undergo polymerisation to form common products such as **polyesters** used for clothing and recyclable polyethylene terephthalate (**PET**) drink bottles. An interesting example of an ester is nitroglycerine (which is somewhat incorrectly named).

11.14 Amides

Amides are derivatives of carboxylic acids in a similar manner to esters. When the OH group is replaced by NH_2, then an **amide** is formed. A special class of amides occurs when the OH group is replaced by NHR. Amides can be produced by the reaction between ammonia and an ester.

The reaction between methyl acetate and ammonia produces **ethanamide** (acetamide) and methanol:

$$CH_3COOCH_3 + NH_3 \rightarrow$$
$$CH_3CONH_2 + CH_3OH$$

$$CH_3C - C \overset{\displaystyle O}{\underset{\displaystyle NH_2}{\Big\|}}$$
acetamide

In the NH_2 group, the electrons are more strongly drawn towards the more electronegative N atoms, leaving the H end with a positive charge and so amides are able to form **hydrogen bonds** with each other as they are attracted to the negatively charged lone pairs on neighbouring O atoms. This confers a relatively high melting point for these compounds. Amides have much the same solubility as comparable esters, with lower amides being soluble in water.

The most notable characteristic of amides are their mechanical strength and are often used, both in nature and industrially, as structural items.

In nature, amides for the basis of links between amino acids in **proteins**.

In industrial use, amides are joined together in long chains to produce polyamides, the most well known of which are **nylon** and **kevlar**.

Lysergic acid diethylamide (**LSD**) is also an amide.

Methanamide	$HCONH_2$
Ethanamide	CH_3CONH_2
Propanamide	$CH_3CH_2CONH_2$

The **amide link** joins amino acids together in biological processes to form **proteins**. **Amino acids** are carboxylic acids with an amine group of the form $H_2NCHRCOOH$. During polymerisation, water is eliminated between two acid groups to be joined together.

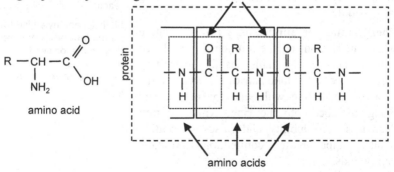

amino acid

11.15 Amines

In an alcohol, the R takes the place of one of the H atoms in the water molecule to give the general formula ROH. Now consider the molecule **ammonia** NH_3.

If one or more H atoms are replaced by an alkyl group, then the resulting compound is called an **amine**. Primary amines RNH_2 have one R substitution, secondary amines R_2NH have two, and tertiary amines R_3N have all three H atoms replaced with alkyl groups. Aliphatic and aromatic structures are found.

Ammonia is a colourless gas with a very sharp odour. Ammonium hydroxide in household cleaning products is a solution of ammonia gas in water where it slightly dissociates into the ammonium NH^{4+} and hydroxide OH^- ions.

Many primary aromatic amines are used as dyes.

Aniline

Ammonia	NH_3
Methylamine	CH_3NH_2
Ethylamine	$CH_3CH_2NH_2$
Propylamine	$CH_3CH_2CH_2NH_2$
Butylamine	$CH_3CH_2CH_2CH_2NH_2$
Aniline	$C_6H_5NH_2$

Just as in the case of ammonia, amines are bases due to the lone electron pair on the N atom. Primary and secondary amines can form hydrogen bonds with each other and all can form **hydrogen bonds** with water, thus making them soluble, the lower gaseous amines being very soluble.

Phenylamine, or **aniline**, is a primary amine where an H atom on the ammonia is replaced with a benzene ring. This has the interesting consequence of the lone pair of electrons on the N atom interacting with, and to some extent being absorbed into the ring, which in turn decreases its solubility and raises its boiling point compared to comparable aromatic compounds.

Amines are important in biological functions and are produced from the breakdown of amino acids in the body. **Amino acids** are carboxylic acids with an amine group of the form $H_2NCHRCOOH$. Amines serve as neurotranmitters. A common natural amine is **histamine** which triggers an immune response. Anti-histamines suppress this activity in severe allergic reactions.

Methylphenethylamine (or **amphetamine**) is a powerful drug that modifies the action of natural neurotransmitters in the brain.

11.16 Polymers

The chemical and physical properties of hydrocarbon compounds and derivatives depend very much on the size of the molecule. Smaller groups such as methyl and ethyl compounds generally have a high solubility, a low boiling point. When these small molecular units, collectively called **monomers,** and joined together into long chains, the molecular size can increase greatly with the production of a variety of useful substances which are called **polymers,** or **plastics.**

For example, in ethylene, the C=C bonds may be successively broken and joined to form larger and larger molecules of polyethylene. This process is usually performed with the help of a catalyst. The resulting compounds are called **addition polymers.** When other atoms or functional groups are involved, polymers of a great variety of properties are possible.

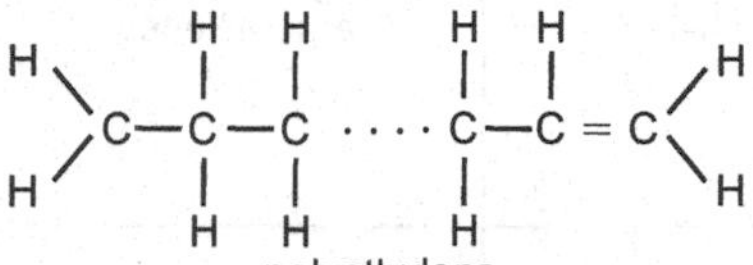

Another class of polymers are **condensation polymers.** In these, chains are formed by the expulsion of a water molecule from between the monomers to be joined. In **Nylon,** amide links are used to join monomers into large chain molecules.

One of the monomers used to make **Nylon** 6,6 is adipic (hexanedioic) acid:

$$HOOC-CH_2-CH_2-CH_2-CH_2-COOH$$

This is used with another monomer, 1,6 diaminehexane:

$$H_2N-CH_2-CH_2-CH_2-CH_2-CH_2-CH_2-NH_2$$

When these two monomers are chained together, water drops out as shown:

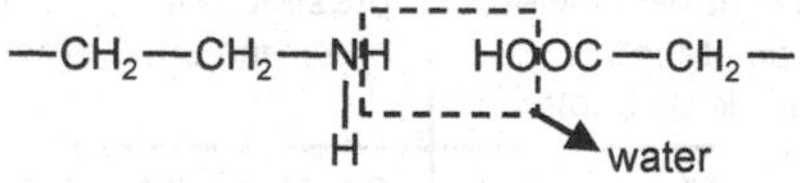

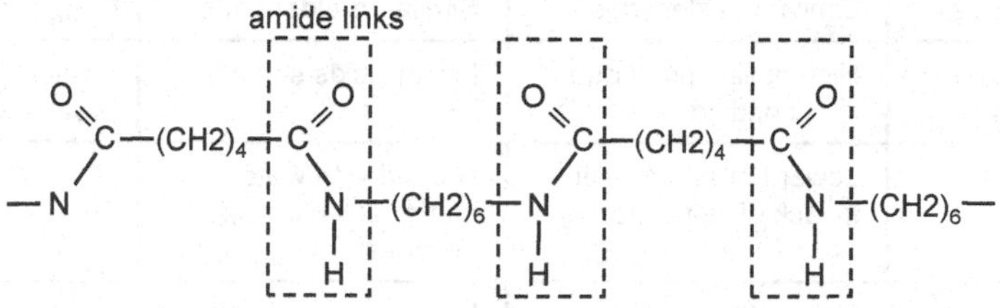

During polymerisation, these molecules continue to chain together, expelling a water molecule each time and forming amide linkages. **Kevlar** is a similar polymer but the amide links join benzene rings together.

11.17 Physical and Chemical Properties

Group	Boiling point	Solubility	Notes
Alkanes	Lower compounds are gases at room temp. Boiling point increases as size of molecule increases.	Do not form H bonds, insoluble in water.	Mainly unreactive with other compounds.
Alkenes	Similar to alkenes.	Similar to alkenes.	Slightly more reactive due to presence of C=C.
Alkynes			More reactive than alkenes due to C≡C.
Alcohols	H bonding causes boiling point to be higher than comparable alkanes.	Lower alcohols are soluble in water. Compounds become less water-like with increasing size.	OH group makes the molecules polar, and able to form H bonds.
Ethers	Cannot form H bonds between themselves so lower boiling point than comparable alcohol.	Can form H bonds with water molecules. Lower ethers soluble in water.	More reactive than comparable alkanes.
Aldehydes	Cannot form H bonds with each other and so generally have a lower boiling point than comparable alcohols.	Lower aldehydes soluble in water due to presence of lone pair electrons on O atom.	
Ketones	Similar to aldehydes.	Similar to aldehydes.	
Carboxylic acids	High boiling point due to H bonding.	Lower acids soluble.	Typically weak acids.
Esters	Lower boiling point due to lack of self-H bonds.	H bond with water confer solubility on lower esters.	
Amines	Reasonably high boiling point but lower than alcohols.	Lower amines soluble.	

12. Biochemistry

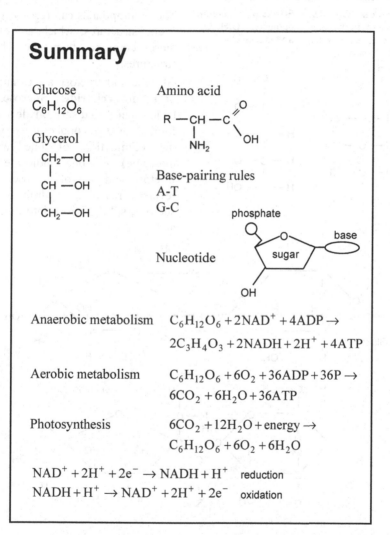

Summary

Glucose
$C_6H_{12}O_6$

Glycerol

CH_2—OH
|
CH —OH
|
CH_2—OH

Amino acid

R —CH —$C \overset{\displaystyle O}{\underset{\displaystyle OH}{\diagup\hspace{-6pt}\diagdown}}$
|
NH_2

Base-pairing rules
A-T
G-C

Nucleotide

phosphate — sugar — base — OH

Anaerobic metabolism

$C_6H_{12}O_6 + 2NAD^+ + 4ADP \rightarrow$
$2C_3H_4O_3 + 2NADH + 2H^+ + 4ATP$

Aerobic metabolism

$C_6H_{12}O_6 + 6O_2 + 36ADP + 36P \rightarrow$
$6CO_2 + 6H_2O + 36ATP$

Photosynthesis

$6CO_2 + 12H_2O + energy \rightarrow$
$C_6H_{12}O_6 + 6O_2 + 6H_2O$

$NAD^+ + 2H^+ + 2e^- \rightarrow NADH + H^+$ reduction
$NADH + H^+ \rightarrow NAD^+ + 2H^+ + 2e^-$ oxidation

12.1 Sugars

Sugars are part of a larger group of compounds called **carbohydrates**. Carbohydrates have the general formula $C_m(H_2O)_n$ and can be thought of as hydrated carbon compounds.

Glucose, $C_6H_{12}O_6$, an important sugar in biochemistry is an aldehyde.

Fructose, an isomer of glucose, $C_6H_{12}O_6$, and is a ketone.

These compounds can readily form cyclic structures, which are often more common than the chain structures.

Glucose and fructose are examples of **monosaccharides**. **Sucrose**, or table sugar, is a **disaccharide** and is formed by the condensation (with the elimination of one water molecule) of one molecule of fructose and one of glucose. The reverse reaction, hydrolysis of sucrose, splits sucrose into glucose and fructose.

Glucose is the basic fuel for the process of **metabolism**. Glucose is recovered from fats and sugars in the diet and stored as **glycogen** in the body until converted back to glucose for use directly by cells.

12.2 Polysaccharides

Cellulose is a straight-chain polysaccharide $(C_6H_{10}O_5)_n$ consisting of thousands of glucose units. It is insoluble in water. The nature of particular forms of cellulose in organisms depends mainly upon the chain length of the molecule.

Hydrogen bonding between parts of the molecule impart strength to the structure.

Cellulose

Starch is similar to cellulose in that it is constructed from glucose monosaccharide units, but has a different orientation of linkage between the glucose elements, with more branching than cellulose. It is generally a mixture of glucose polymers of varying lengths, the shorter constituents (of molecular weight ≈4000) being water soluble.

Starch typically contains a mixture of linear (soluble) amylose and branched (relatively insoluble) amylopectin molecules. Complete hydrolysis gives glucose. Partial hydrolysis results in various starch sugars called dextrins.

Starch is broken down into glucose in the body by enzymes, (chiefly amylase, which is a constituent of saliva and pancreatic juice).

Glycogen is similar to the amylopectin form of starch in structure and is found in the liver and muscles of animals as an energy store. It is synthesised from glucose from digestion by enzymes in the liver where it is stored until released as **glucose** into the bloodstream as needed.

Cellulose acts as the main structural element of plants and forms the cell wall that gives plants their rigidity. Wood is about 50% cellulose. Mammals generally are incapable of breaking down cellulose to glucose although cows and sheep, and other ruminants, have the necessary bacteria in their digestive system that perform this function. Insects (e.g. termites) can also digest cellulose.

Starch is used as an energy store in plants. The glucose produced by photosynthesis is stored in plant tissue as starch until it is needed by the plant. Starch in plants, when eaten as food, is an important source of glucose for animals.

12.3 Lipids

Fats and oils are **esters**. Esters are derivatives of **carboxylic acids** where the H on the OH group is replaced by R so that the general formula is RCOOR'. Esters are usually derived from the reaction between acids and alcohols with the elimination of a water molecule.

Animal and plant fats are triple esters, or **triglycerides**, of **glycerol (glycerin)** $C_3H_8O_3$. Each of the three available OH groups in glycerol is available for combination with three molecules of **fatty acid**. Fatty acids typically have from 12 to 20 carbon atoms.

$$CH_2-OH$$
$$|$$
$$CH\ -OH$$
$$|$$
$$CH_2-OH$$

Glycerol
(propane 1,2,3 triol)

Fat in the body acts as a storage place for glucose. When fat reserves are called upon, the fats are converted back into glycerol and **fatty acids**. The glycerol in turn is converted into glucose by the liver.

Fats, which are solid at room temperature, are made from **saturated fatty acids**, while **oils**, which are liquid at room temperature, are formed mainly from **unsaturated fatty acids**. Fats and oils from plants and animals typically contain a mixture of different types of esters, two or three fatty acids attached to a glycerol molecule.

Fats in the diet are also a source of **linolenic** $C_{17}H_{29}CO_2H$ and **linoleic** $C_{17}H_{31}CO_2H$ unsaturated fatty acids which are essential in animal dietary intake because they are unable to be synthesised by the body directly and are required for metabolism into a variety of other acids required by various bodily functions.

Linolenic acid has the general formula

$$CH_3(CH_2)_4CH=CHCH_2CH=CH(CH_2)_7COOH$$
$$\uparrow$$

and is called an omega-6 fatty acid because the first C=C double bond is on the sixth carbon atom from the (left-hand) end. **Linoleic acid** is am omega-3 acid. (Note: normally numbering of atom positions is from the right.)

$$CH_3CH_2CH=CHCH_2CH=CHCH_2CH_2=CH(CH_2)_7COOH$$
$$\uparrow \qquad\qquad\qquad\qquad\qquad\qquad Linoleic\ acid$$

Fats also serve a structural role in the body by providing heat insulation and a barrier against mechanical shock for organs.

Fats, along with oils and waxes, are one example of a broader group of **lipids**. Compared to carbohydrates, lipids generally contain a smaller proportion of O atoms. Lipids are relatively insoluble in water.

When fats hydrolyse, or react with water, in an alkaline solution, glycerol is formed along with the metal salt of the carboxylic acid. Such a reaction is called **saponification**, from which soaps (e.g. sodium stearate) is formed.

12.4 Proteins

Proteins are amino acids joined together by **amide** (or **peptide**) links and are classified as **polyamides**. The arrangement, or sequence, of amino acids within the protein structure determines the function of the protein. A typical protein may contain 100 or more amino acids.

Amino acids with the amine group attached to the same C atom as the COOH group are called **alpha-amino acids** and are the most important in biochemistry.

$$R - CH - C \begin{smallmatrix} \diagup O \\ \diagdown OH \end{smallmatrix}$$
$$| \quad\quad NH_2$$

amino acid

The mechanical structure of proteins (i.e. the sequence of amino acids and hydrogen bonds) has great significance for living organisms. Proteins form the basis of skin, hair, muscle, tendons, and other tissues in the body.

The twenty or so amino acids found in proteins differ in the makeup of the R residual. In **glycine**, R is just an H atom, while in phenylaniline, R is a ring structure.

On one end of the chain, there is a free NH_2 group; this is the N terminal. At the other end of the chain, there is a COOH group, the C terminal. In between, water molecules have been eliminated to leave amino acid units, or residues.

Some proteins are enzymes and act as catalysts for chemical reactions which would otherwise proceed too slowly for use in organisms.

$$NH_2 - \underset{\underset{H}{|}}{\overset{\overset{R}{|}}{C}} - \underset{\underset{H}{|}}{N} - \overset{\overset{O}{||}}{C} - \underset{\underset{H}{|}}{\overset{\overset{R}{|}}{C}} - \ldots - \underset{\underset{H}{|}}{N} - \overset{\overset{O}{||}}{C} - \underset{\underset{H}{|}}{\overset{\overset{R}{|}}{C}} - COOH$$

A typical **polypeptide** protein consists of between 100 and 500 or more amino acid residues. The sequence of amino acids is the **primary structure** of the molecule. The long chains of proteins themselves have a **secondary structure**, usually in the form of a helix or spiral. **Hydrogen bonds** between different amino acid groups are responsible for this structure. Often, there is more than one polypeptide chain present and they exist as intertwined helixes held together by hydrogen or ionic bonds leading to a **tertiary structure**.

When a protein undergoes complete hydrolysis, the amino acids are recovered.

The secondary structure of proteins is stable within narrow temperature and pH ranges. When the secondary structure of a protein is disrupted, either by heat (such as by cooking), or immersion in acids or alkalis, the protein is said to be denatured and the physical properties change markedly.

12.5 Nucleic Acids

Carbohydrates, lipids and proteins make up the bulk of living organisms. A fourth major class of compounds comprise the nucleic acids **DNA** and **RNA**. Nucleic acids are long-chain polymers that consist of smaller units, **nucleotides**, joined together.

In DNA, there are four types of nucleotide. Each nucleotide consists of a 5-carbon sugar (**deoxyribose**), with an attached phosphate, and a nitrogen base. Each type is distinguished by the identity of the nitrogenous bases.

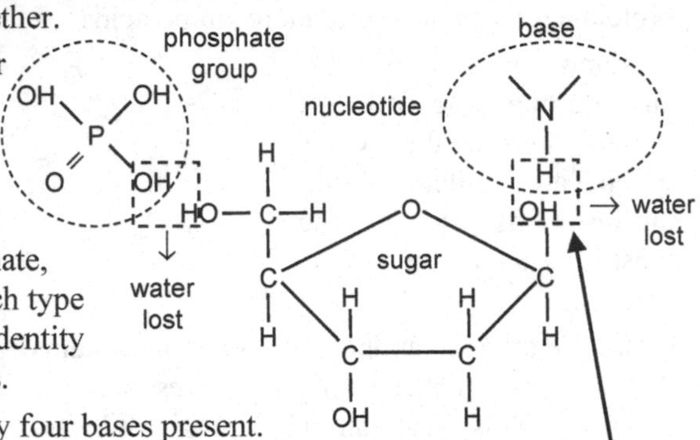

In DNA, there are only four bases present.

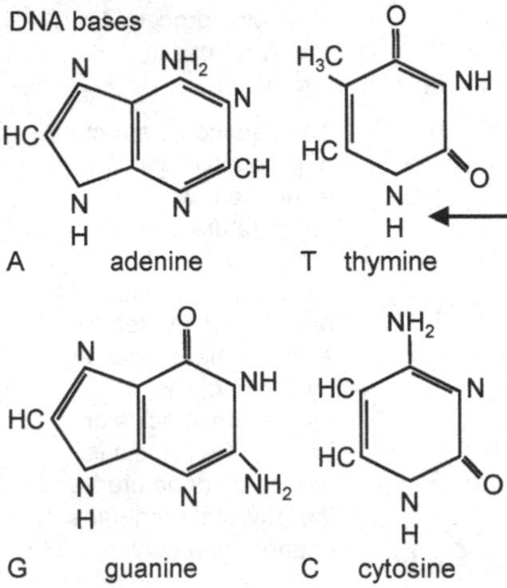

phosphate group

nucleotide

base

water lost

water lost

sugar

DNA bases

A adenine

T thymine

G guanine

C cytosine

The NH group on the base combines with the OH group on the sugar and a H_2O is lost.

The primary structure of DNA is a sequence of nucleotides bonded together as long chains. The phosphate group of one nucleotide bonds with the sugar of another (losing a molecule of water in the process).

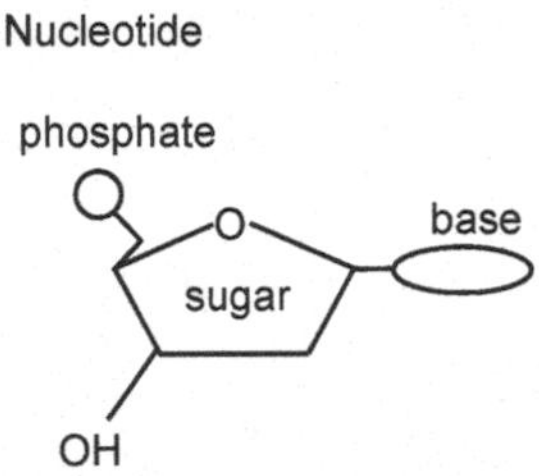

Nucleotide

phosphate

sugar

base

OH

A shorthand way of writing these groups is to just show the covalent bonds between the groups, and leave out the C and H atoms.

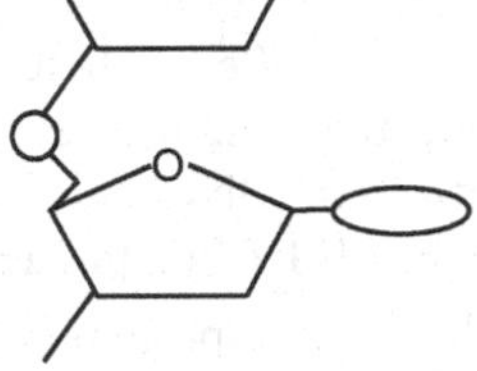

12.6 DNA

The primary structure of DNA is essentially the sequence of bases in the nucleotides in the chain. However, DNA does not ordinarily exist as a single chain, but is paired with an opposite chain so that the bases are paired and held together by **hydrogen bonds**.

The bases can only pair according to certain rules: A is always paired with T by a double hydrogen bond. G is always paired with C by a triple hydrogen bond.

The ladder-like structure is further characterised by being coiled up in a spiral, each chain with bases pointing inwards and bonding with the corresponding base on the other – the so-called **double-helix**. The nucleotides, identified by their bases, can appear in any particular order, but the sequence on one chain has to be reflected in the other by virtue of the **base-pairing rules** A-T and G-C.

Sequences of nucleotides fall into functional groups called **genes**, which in turn lie along a single large DNA molecule called a **chromosome**. In most cells, chromosomes occur in duplicated pairs. When cell division by **mitosis** occurs, the DNA molecules are replicated by rupturing the hydrogen bonds between the two chains and then each chain forming bonds with new nucleotides according to base-pairing rules. When cell division is complete, the two new cells each contain paired chromosomes of the same genetic sequence as the parent cell.

Although there are only four bases in human DNA, the molecule typically consists of about 1500 nucleotides in the chain, allowing billions of combinations of base sequences possible in a single molecule.

RNA is similar to DNA but consists of ribose as the sugar and a single chain structure with the base **uracil** rather than thymine. There are different types of RNA; some act as messengers, carrying the DNA template to the site of protein synthesis (**ribosomes**) in the cell, while others transport amino acids to the ribosomes.

12.7 Enzymes

The many chemical reactions which take place in an organism to sustain life are collectively called **metabolism**. Typical reactions are **condensation** and **hydrolysis** reactions:

sugars		polysaccharides
	Condensation	(cellulose, starch)
fatty acids +	→	fats
glycerol		
	Hydrolysis	
amino acids	←	proteins

These reactions occur at temperatures of about 37°C under atmospheric pressure. The reaction rates would be too slow unless reaction pathways were altered by use of **catalysts**. Nearly every biological chemical reaction involves the use of a specific biological catalyst called an **enzyme**.

Enzymes are very large protein molecules of a specific shape. The enzyme molecule is typically much larger than the molecules actually involved in the chemical reaction. The reaction molecule (or molecules in the case of a condensation reaction), called the substrate, attaches itself to an **active site** on the enzyme molecule.

An **activated complex** is formed, and the reaction proceeds along a different pathway than would normally occur in the laboratory.

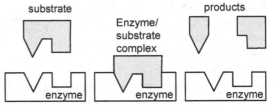

When the reaction is complete, the products detach from the enzyme and the enzyme is available to catalyse another reaction.

Enzymes lower the **activation energy** by increasing the collision frequency of molecules by virtue of bringing the molecules together spatially in an optimum manner, or altering the bond energies involved in the reaction by forming temporary bonds on the reacting molecules, and in some cases, applying mechanical stress to the reacting molecules to facilitate contact.

Enzyme inhibitors are used within the organism to control the activity of enzymes by masking or blocking the active sites on the enzyme to regulate reaction rates as a whole.

The importance of enzymes cannot be over-stated. In a living cell, the reagents of all the chemical reactions are not stored in isolated bottles and mixed when needed as in a laboratory, but exist all together in solution. The process of chemical reactions between these mixed reagents is orchestrated by the action of enzymes.

12.8 ATP

Organisms need nutrients to stay alive. Nutrients comprise carbohydrates, proteins and fats and, in animals, are usually ingested as food. Nutrients provide a source of energy, essential amino acids and essential fatty acids. Vitamins and other trace elements are also required for the proper functioning of enzymes.

Metabolism is the oxidation of glucose to carbon dioxide, water and energy:

$$C_6H_{12}O_6 + 6O_2 + 36ADP + 36P \rightarrow 6CO_2 + 6H_2O + 36ATP$$

The energy released by the oxidation process is stored as chemical potential energy in the formation of **ATP** molecules from **ADP** molecules. **Adenosinediphosphate** consists of a molecule of adenine, ribose, and two phosphate groups:

Energy stored as phosphate bonds via the conversion of ADP (adenosinediphosphate) to ATP (adenosinetriphosphate)

$$ADP + P + energy \leftrightarrow ATP + H_2O$$

adenine   ribose   phosphates

When energy is used to bind a third phosphate group to the end of the molecule, **adenosinetriphosphate** is formed. This is an endothermic process and is called **phosphorylation**. The energy used to create ATP is stored in the phosphate bond. Energy-rich ATP molecules are then hydrolysed to release their energy when used directly in other biological processes such as muscle action, protein synthesis and nerve impulse transmission.

Another important molecule involved in metabolism is the co-enzyme NAD (**nicotinamide adenine dinucleotide phosphate**). NAD takes part in oxidation-reduction reactions in metabolism by losing and accepting electrons ultimately necessary for conversion of ADP to ATP. The redox reactions are

$$NAD^+ + 2H^+ + 2e^- \rightarrow NADH + H^+ \quad \text{reduction}$$

Reduction can occur by adding hydrogen.

$$NADH + H^+ \rightarrow NAD^+ + 2H^+ + 2e^- \quad \text{oxidation}$$

Oxidation can occur by removing hydrogen.

When NAD^+ is reduced, it stores energy by storing an excited electron in the form of NADH and a H^+ ion. When NADH+H^+ is oxidised, it releases energy, usually to form ATP from ADP.

12.9 Anaerobic Metabolism

Anabolic processes build up low-energy reactants into high-energy products. Photosynthesis in plants is an anabolic process. **Catabolic** processes break down high-energy reactants to low-energy products, releasing energy in doing so. This energy is transferred to **ATP**.

Anaerobic metabolism involves two steps: (i) the breakdown of glucose from a **carbohydrate** into **pyruvic acid**:

$$C_6H_{12}O_6 + 2NAD^+ + 4ADP \rightarrow 2C_3H_4O_3 + 2NADH + 2H^+ + 4ATP$$

and (ii) the fermentation of pyruvic acid into either CO_2 and ethanol, or lactic acid. The ultimate aim is to produce ATP from ADP.

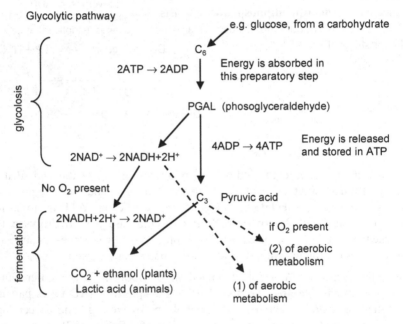

Anaerobic metabolism occurs in all cells. It does result in a particularly high yield of ATP since the products, ethanol, lactic acid and pyruvic acid, still contain a substantial portion of the original energy from the glucose. While anaerobic metabolism can provide ATP for energy use (in muscles) when O_2 is in short supply, or none at all, it also provides a source of **pyruvic acid** for use in **aerobic respiration**.

12.10 Aerobic Metabolism

Aerobic metabolism, or **respiration**, is the process by which glycerol is oxidised with the production of ATP molecules from ADP molecules in the presence of molecular oxygen:

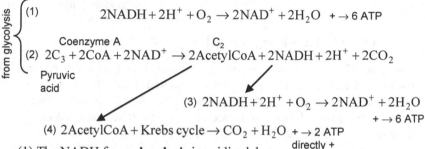

from glycolysis

(1) $2NADH + 2H^+ + O_2 \rightarrow 2NAD^+ + 2H_2O$ + → 6 ATP

Coenzyme A C_2

(2) $2C_3 + 2CoA + 2NAD^+ \rightarrow 2AcetylCoA + 2NADH + 2H^+ + 2CO_2$

Pyruvic acid

(3) $2NADH + 2H^+ + O_2 \rightarrow 2NAD^+ + 2H_2O$

+ → 6 ATP

(4) $2AcetylCoA + Krebs\ cycle \rightarrow CO_2 + H_2O$ + → 2 ATP

(1) The NADH from **glycolysis** is oxidised, by the addition of O_2, to produce six ATP.

directly + NADH+H$^+$

(2) **Pyruvic acid** C_3 is oxidised to an activated form of acetic acid C_2. NAD$^+$ is reduced to NADH+H$^+$.

(3) NADH+H$^+$ from (2) is oxidised by O_2, to form six ATP.

(4) **AcetylCoA** enters the **Krebs cycle** (or **citric acid cycle**). This consists of multiple reactions, the net effect of which is to form two molecules of ATP plus energy-rich NADH.

During the Krebs cycle, six molecules of NADH+H$^+$ and two molecules of a related compound, FADH, are formed. These electron-carrier molecules are energy-rich and can be further oxidised as in (1) and (3) which require O_2. The net result of (4) is the production of twenty two ATP molecules. **Anaerobic metabolism** produces two molecules of ATP, which added to the thirty four ATP molecules from aerobic metabolism makes thirty six molecules of ATP from the complete oxidation of one molecule of **glucose**.

Fats and **proteins** are also sources of ATP, by virtue of their breakdown into components which can be inserted into glycolytic and aerobic pathways.

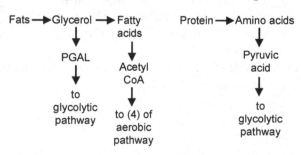

Fats → Glycerol → Fatty acids

PGAL

Acetyl CoA

to glycolytic pathway

to (4) of aerobic pathway

Protein → Amino acids

Pyruvic acid

to glycolytic pathway

12.11 Cyclic Photophosphorylation

Ultimately, the energy contained within glucose used in animal metabolism comes from **photosynthesis** in plants. The overall reaction in plants, for the production of glucose, is

Note that O_2 is a product of this reaction.

$$\text{light} \downarrow$$

$$6CO_2 + 12H_2O + \text{Energy} \rightarrow C_6H_{12}O_6 + 6O_2 + 6H_2O$$

Chlorophyll is the unique ingredient of plants that captures photons to promote an electron of low energy into an excited state. The excited electron is passed from one transfer molecule to another, losing energy at each step, until it returns to a chlorophyll molecule in an unexcited state. At each step, the energy lost is used to convert ADP to ATP:

$$\text{ADP} + \text{P} + \text{Energy} \leftrightarrow \text{ATP} + H_2O$$

As the excited electron is passed from molecule to molecule, *some* of the energy is used to create ATP. The process is not 100% efficient.

Because electrons are donated by chlorophyll from the high-energy state and then to acceptance in a low-energy state, this synthesis of ATP is called **cyclic photophosphorylation**. The ATP, essentially created from ADP by the energy of sunlight, is ultimately used to create high-energy glucose (which is stored in the plant as starch) from low-energy CO_2. Animals eat the plants and convert starch into glucose, which is then used to create ATP for their own metabolic processes.

PQ, Cyt and PC are electron carrier molecules that undergo reduction (when they accept the electron) and oxidation (when they donate the electron).

In this mode of photophosphorylation, electrons are transported around in a cycle; no oxygen or NADH+H$^+$ is produced.

12.12 Non-Cyclic Photophosphorylation

Another way of producing ATP used by plants has a beginning similar to cyclic photophosphorylation, but employs a more complicated path for extraction of energy from excited electrons. This non-cyclic photophosphorylation is used by modern green plants :

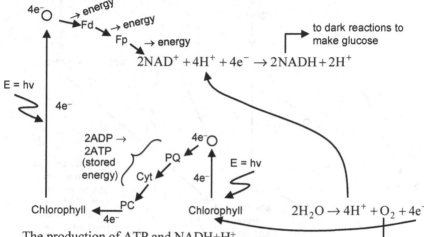

The production of ATP and NADH+H$^+$ involves two **light reactions**. Electron deficiency in one chlorophyll is balanced by different electrons from those from the water dissociation.

Note that O$_2$ is a product of this reaction

Additional reactions take place in which the energy obtained from the light reactions is converted into carbohydrates for storage. That is, CO$_2$ is reduced, and (if the energy is coming from NADH+H$^+$) the NDAH+H$^+$ is oxidised with the net result being glucose. These are known as **dark reactions** since no light is required.

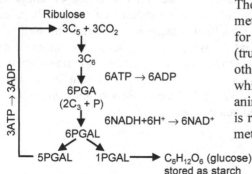

The resulting glucose is used in metabolism by the plant cells for making cellulose/starch (trunks and branches, etc) and other sugars (such as in fruit), which may in turn be eaten by animals from which the glucose is recovered and used in animal metabolism.

12.13 Metabolism

Summary:

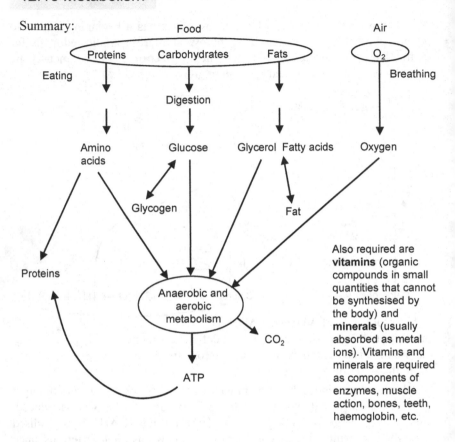

ATP is used to provide the energy source for conversion of amino acids into proteins required by the body for the production of skin, hair, cells, enzymes, muscles, nerves, signalling, anaerobic respiration, as well as for energising DNA replication. It also provides energy for muscle action, vision, brain activity, and nearly all biological energy transformations. Glycogen and fats can be stored in the body and sent back into the metabolism pathway as glucose and glycerol for oxidation into CO_2 and energy when there is no food being digested.

Index

Printed in the United States
by Baker & Taylor Publisher Services